BG Chemie

Springer

Berlin
Heidelberg
New York
Barcelona
Budapest
Hong Kong
London
Milan
Paris
Santa Clara
Singapore
Tokyo

Toxicological Evaluations

10

Potential Health Hazards of Existing Chemicals

Springer

Berufsgenossenschaft
der chemischen Industrie
Kurfürsten Anlage 62

D 69115 Heidelberg

Library of Congress Cataloging-in-Publication Data (Revised for vol. 10)
Toxicological evaluations.
Includes bibliographical references and indexes.
1. Toxicity testing. I. Berufsgenossenschaft der Chemischen Industrie. [DNLM: 1 Hazardous Substances Toxicity. WA 455 T7549
RA1199.T678; 1990; 615.9'07; 90-10020
ISBN-13: 978-3-642-80076-4 e-ISBN-13: 978-3-642-80074-0
DOI: 10.1007/ 978-3-642-80074-0

Softcover reprint of the hardcover 1st edition 1996

Production: PRODUserv Springer Produktions-Gesellschaft, Berlin
Typesetting: Dataconversion by Martin Schenk, Lewis & Leins GmbH, Berlin

SPIN 10476350 51/3020 - 5 4 3 2 1 0 - Printed on acid-free paper

Preface

As part of its "Programme for the prevention of health hazards caused by industrial substances", the Berufsgenossenschaft der chemischen Industrie (BG Chemie, Employment Accident Insurance Fund of the Chemical Industry) began in 1977 to investigate the toxicity of those chemicals which are widely used, have many different applications and are suspected of being dangerous to health, in particular of having long-term effects. The investigations consist of a literature search and – depending on the results – commissions of experimental studies. It is hoped by means of this testing to close gaps in our knowledge and to increase the scientific validity of the required risk assessments. The results of the toxicological investigations carried out by BG Chemie, and the resulting substance assessments have been published in German since 1987 in the form of 189 "Toxikologische Bewertungen" ("Toxicological Evaluations") up to now.

In order to make this useful information internationally available, BG Chemie began in October 1990 to publish them as a book series in English, of which the tenth volume (containing 11 individual evaluations) is presented here. Therefore for 139 existing chemicals, "Toxicological Evaluations" are available in English at the moment, a further 54 are in preparation and will be published soon.

Because of the short time between publishing volume 4 - 10, printing of an "Introduction" (consisting of a general overview of the programme, lists with names of people involved as well as substances under investigation) has been abandoned in this volume and will be included in volume 11 again. If more detailed information is required about the ongoing work, see volume 1 or 4 or contact BG Chemie at first hand.

BG Chemie hopes that, for many people working in the chemical industry, this information will be of practical help in assessing hazards to health at the workplace.

Dr. Radek

Contents

Dicyclopentadiene

1. Summary and assessment

After oral administration to mice, rats and dogs, ^{14}C-labelled dicyclopentadiene is mainly excreted in the urine (75 % in the mouse, 53.7 % in the rat, 64.5 % in the dog) with 2-phase elimination kinetics. Half-lives of 4 h in the mouse and 10 h in the dog have been calculated for the rapid phase (no exact calculation in the rat due to insufficient data), while the corresponding values for the slow phase were 18 h in the mouse and rat and 27 h in the dog. In cows, dicyclopentadiene is eliminated mainly in the urine (80 % in 48 h).

On acute exposure, dicyclopentadiene is moderately toxic after oral administration, and of low toxicity on dermal application. The LD_{50} for dicyclopentadiene in the rat is between 347 and 820 mg/kg body weight after oral administration, and >2000 mg/kg body weight after dermal application, while that in the mouse is between 190 and 250 mg/kg body weight after oral administration and that in the rabbit is >2000 mg/kg body weight following dermal application. Signs of toxicity observed include ataxia, reduced movement, exhaustion, tremors and spasms. The 4-hour LC_{50} in the rat is 284 to 385 ppm (equivalent to 1534 to 2079 mg/m^3), and the substance is thus toxic on inhalation.

Repeated administration of 64 mg dicyclopentadiene/kg body weight/day in the diet for 14 days is tolerated by dogs without overt signs of toxicity. In minks, 754 mg/kg body weight/day administered in the diet for 21 days caused changes to haematological parameters and a marked loss in body fat. On administration of 40 or 200 mg dicyclopentadiene/kg body weight/day to rats by gavage for 28 days, body weight gain was inhibited and changes in liver, kidney, adrenal gland and thymus weights were seen, as well as histological effects on the liver and adrenal glands. The no-observed-effect level was given as 8 mg/kg body weight/day. These findings are contrary to the results of subchronic feeding studies in rats and mice, in which no effects occurred at doses of up to 198 and 88 mg/kg body weight, respectively. Inhalation of dicyclopentadiene vapour 10 times (7 h/day) can be lethal in rats at concentrations of 250 ppm (equivalent to 1350 mg/m^3) and above and in mice at 72 ppm (equivalent to 389 mg/m^3) and above. Bleeding in the lungs, gut and thymus have been observed macroscopically in rats.

The substance is irritating to the skin and eyes of the rabbit.

Dicyclopentadiene is not a skin sensitizer in guinea pigs.

No effects were detected after administration of dicyclopentadiene in the diet for 90 days up to the highest tested doses of 198 mg/kg body weight/day in rats and 88 mg/kg body weight/day in mice. Inhalation of dicyclopentadiene vapour for 13 weeks caused histological effects on the kidneys (α2u-globulin nephropathy) in male rats at a concentration of 5.1 ppm (equivalent to 27.5 mg/m^3) and above. The a2u-globulin nephropathy is specific to male rats and is not relevant to risk assessment in man. In dogs, no histopathological effects have been established up to concentrations of 32.4 ppm (equivalent to 175 mg/m^3), although minor changes in clinical chemistry parameters were evident. No-effect levels of 1 ppm for the rat and between 8.9 and 23.5 ppm for the dog were given.

The substance shows no mutagenic potential in the Salmonella/microsome test.

No indications of carcinogenic effects have been found in the rat following intramuscular injection of dicyclopentadiene (study conditions not further described).

Dicyclopentadiene shows neither maternal nor foetal toxicity in rats and minks and is not teratogenic at doses of up to 170 mg/kg body weight/day. No treatment-related effects were seen in a 3-generation study in rats involving the administration of the substance in the diet at doses of up to 50 mg/kg body weight/day.

In volunteers, inhalation of 1 ppm dicyclopentadiene for 30 minutes caused mild eye and throat irritation, as well as a reduced sense of smell. The olfactory threshold in man is between 0.003 and 0.5 ppm.

A carcinogenicity study with dicyclopentadiene is planned as part of the National Toxicology Program (NTP).

2. Name of substance

2.1	Usual name	Dicyclopentadiene
2.2	IUPAC-name	Tricyclo[5.2.1.0^{2,6}]deca-3,8-diene
2.3	CAS-No.	77–73–6
2.4	EINECS-No.	201–052–9

3. Synonyms, common and trade names

Bicyclopentadiene
CP-dimer
Cyclopentadiene dimer
1,3-Cyclopentadiene dimer
DCP
DCPD
4,7-Methano-1H-indene,3a,
4,7,7a-tetrahydro-
3a, 4, 7, 7a-Tetrahydro-
4,7-methanoindene
Dicyclopentadien

4. Structural and molecular formulae

4.1 Structural formula

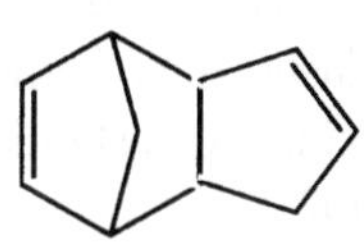

4.2 Molecular formula $C_{10}H_{12}$

5. Physical and chemical properties

5.1	Molecular mass, g/mol	132.20	
5.2	Melting point, °C	32.5	(Dow, 1992)
5.3	Boiling point, °C	172.8 (at 1013 hPa)	(Dow, 1992)
5.4	Vapour pressure, hPa	4.0 (at 20 °C)	(Griesbaum et al., 1987)
5.5	Density, g/cm^3	1.071 (at 20 °C) 0.977 (at 35 °C)	(NTP, 1990) (Dow, 1992)
5.6	Solubility in water	40 mg/l	(Dow, 1992)
5.7	Solubility in organic solvents	Soluble in ethanol	(Sax and Lewis, 1986)
5.8	Solubility in fat	No information available	
5.9	pH value	–	
5.10	Conversion factor	1 ml/m^3 (ppm) ≙ 5.4 mg/m^3 1 mg/m^3 ≙ 0.19 ml/m^3 (ppm)	(at 1013 hPa and 25°C)

6. Uses

As an intermediate in the manufacture of alkyd resins, synthetic rubber, perfume ingredients, plasticizers, flame retardants, pharmaceuticals, paints, lubricants, antioxidants, pesticides and other organic chemicals (Scherb, 1975; Dow, 1992).

7. Experimental results

7.1 Toxicokinetics and metabolism

In order to study the toxicokinetics of dicyclopentadiene, 3 different studies were carried out on the species listed below.

- 24 male Swiss-Webster mice, 20 to 30 g
  40 mg ^{14}C-dicyclopentadiene/kg body weight orally
- 12 male Sprague-Dawley rats, 180 to 280 g
  100 mg ^{14}C-dicyclopentadiene/kg body weight orally
- 5 male beagle dogs, 7.6 to 8.9 kg
  100 mg ^{14}C-dicyclopentadiene/kg body weight orally

3 mice were killed each time after 15 minutes and after 1, 2, 4, 6, 24, 48 and 72 h, while 2 rats were killed each time 2, 4, 6, 24, 48 and 72 h after administration. Blood samples were taken from the dogs after 30 minutes and after 1, 2, 4, 6, 10 and 24 h. Maximum concentrations of 11.36 and 39.95 µg ^{14}C-dicyclopentadiene/ml plasma were found in the mice and the dogs, respectively after 2 h, while in the rats, a maximum concentration of 23.28 µg/ml was determined 6 h after administration. The authors determined a 2-phase elimination curve for all 3 species. Half-lives of 4 h in the mouse and 10 h in the dog were found for the rapid phase. Not enough values were available in the rats to enable an exact calculation. For the slow phase, the authors calculated half-lives of 18 h in the mouse, 27 h in the dog and 18 h in the rat. After 1 to 2 h, the radioactivity was distributed throughout the body in all 3 species. The highest values were found in the urinary bladder (without contents), in the bile and in the body fat in the mouse, in the bile and gall bladder (without contents) in the dog, and in the fat, the adrenal glands and the urinary bladder (without contents) in the rat. Within 24 h, all 3 species eliminated a large proportion of the radioactivity in the urine and the faeces.

Table 1. Elimination of dicyclopentadiene

	Urine	Faeces
Mouse	75.0 %	10 %
Rat	53.7 %	15 %
Dog	64.5 %	4 %

In the urine of the mouse and the dog, there were 2 metabolites in each case that could not be identified by thin layer chromatography, while in the rat there were 3 metabolites (Litton Bionetics, 1976; Hart and Dacre, 1978).

A further toxicokinetic study was carried out in a Jersey cow. The animal was given 10 mg dicyclopentadiene/kg body weight/day orally for 5 days, and on the sixth day was given a final dose of 4.0×10^8 dpm ^{14}C-labelled dicyclopentadiene (equivalent to 10 mg dicyclopentadiene/kg body weight). A maximum concentration of 290 dpm/g (equivalent to 2.1 µg dicyclopentadiene/g) was found in whole blood after 2 h. Within 48 h, the cow had eliminated 80.1 % of the radioactivity in the urine, 3.7 % in the faeces and 0.07 % in the milk. No radioactivity was detectable 96 h after administration in the brain, fat, gall bladder, heart, kidneys, liver, muscle tissue, ovaries, lungs, adrenal glands, skin, spleen, urinary bladder or udder (detection limit in the blood: 20 dpm/g (equivalent to 0.15 µg dicyclopentadiene/g); data not provided for detection limit in the tissues). The authors of this study were convinced that the exposure of livestock to small amounts of dicyclopentadiene does not lead to perceptible contamination of the edible flesh or of the milk (Ivie and Oehler, 1980).

7.2 Acute and subacute toxicity

Acute toxicity

The LD_{50} for dicyclopentadiene in the rat was between 347 and 1400 mg/kg body weight after oral administration, 300 mg/kg body weight after intraperitoneal injection and >2000 mg/kg body weight after dermal application, while in the mouse it was between 190 and 1200 mg/kg body weight after oral administration (see Table 2)

Acute signs of toxicity in the mouse and the rat included consistent ataxia, reduced motility, exhaustion, tremors and convulsions. Rats and mice that died during the observation period showed hyperaemia of the lungs. Additional effects seen in the mice included the accumulation of yellowish fluid in the gastro-intestinal tract, dilation of the bladder and effects on the liver and spleen. The authors pointed out that some of the observed symptoms could have been a consequence of autolysis. Post-mortems at the end of the 14-day observation period did not reveal any macroscopically visible adverse effects in either the mice or the rats (Rosenblatt et al., 1975; Litton Bionetics, 1976).

The acute signs of toxicity in mink were characterised by hyperactivity, the production of shrill noises, dyspnoea, diarrhoea, opistho-

Table 2. Data on the acute toxicity of dicyclopentadiene

Species	No. of animals/ dose	Sex	Route	Observation period	LD_{50} (mg/kg body weight)	Reference
Rat (Carworth–Wistar)	–	male	oral	14 days	0.41 ml/kg ($\triangleq$400 mg/kg)	Smyth et al., 1962
Rat	–	male	oral	–	435	Rosenblatt et al., 1975
Rat (Sprague–Dawley)	10	male	oral	14 days	520	Litton Bionetics, 1976; Hart and Dacre, 1978
Rat (Carworth–Wistar)	–	male	oral	14 days	820	Smyth et al., 1954
Rat (Sprague–Dawley)	–	male female	oral	14 days	590	Safepharm, 1989 e
Rat (Sprague–Dawley)	10	female	oral	14 days	378	Litton Bionetics, 1976; Hart and Dacre, 1978
Rat	–	female	oral	–	396	Rosenblatt et al., 1975
Rat (Albino)	–	–	oral	14 days	0.353 ml/kg ($\triangleq$347 mg/kg)	Kinkead et al., 1971
Rat	–	–	oral	–	1400	Prokop'eva and Buzjun, 1988
Rat (Sprague–Dawley)	5	male female	dermal	14 days	>2000	Safepharm, 1989 a
Rat	–	–	dermal	–	>2000	Prokop'eva and Buzjun, 1988

Table 2. (forts.)

Species	No. of animals/ dose	Sex	Route	Observation period	LD_{50} (mg/kg body weight)	Reference
Rat	–	–	i.p.	14 days	0.31 ml/kg (≙300 mg/kg)	Kinkead et al., 1971
Rat	–	–	i.p.	–	200	RTECS, 1992
Rat	–	–	i.p.	–	1075	Prokop'eva and Buzjun, 1988
Mouse (Swiss Webster)	10	male	oral	14 days	190	Litton Bionetics,1976; Hart and Dacre, 1978
Mouse (Swiss Webster)	10	female	oral	14 days	250	Litton Bionetics, 1976; Hart and Dacre, 1978
Mouse	–	–	oral	–	1200	Prokop'eva and Buzjun, 1988
Mouse	–	–	i.p.	–	200	RTECS, 1992
Mouse	–	–	intra–tracheal	–	74	Prokop'eva and Buzjun, 1988
Rabbit	4	male	dermal (24 hours)	14 days	4.46 ml/kg (≙4.38 g/kg)	Smyth et al., 1962
Rabbit	4	male	dermal (24 hours)	14 days	6.72 ml/kg (≙6.60 g/kg)	Smyth et al., 1954
Mink	2–4	male	oral	14 days	>1000	Aulerich et al., 1979
Mink	1	female	i.p.	–	at 960 1/1 died within a few minutes	Aulerich et al., 1979
Calf	4	male female	oral	14 days	ca. 1000	Cysewski et al., 1981

– not specified

tonus, convulsions, vomiting and paresis of the hind legs. The animals returned to normal within 1 to 1.5 h. A dose of more than 960 mg/kg body weight (solvent unspecified) administered by intraperitoneal injection was lethal. No gross pathological effects were evident at post-mortem in the animals that died during the observation period or those killed at the end of the study (Aulerich et al., 1979).

A single oral dose of 250, 500, 1000 or 2000 mg dicyclopentadiene/kg body weight given to 8- to 10-week-old calves (4/group) led to intolerance reactions in all groups. Dose-dependent ataxia and increased salivation occurred from 250 mg/kg body weight, with tonic-clonic spasms from 500 mg/kg. The two higher doses were in the lethal range: at 1000 mg/kg body weight, 1 of the 4 treated calves died after 3 days, while 2000 mg/kg body weight caused the deaths of all 4 animals within 6 days. Beginning with the intermediate doses (500 to 1000 mg/kg), dicyclopentadiene caused dose-dependent increases in the activity of creatine-phosphokinase, aspartate aminotransferase and alanine aminotransferase as well as increases in leukocytes, erythrocytes, haemoglobin and haematocrit values. In the surviving calves, these parameters returned to normal within 14 days. No histopathological changes were noted on microscopic examination of the brain, spinal cord, lungs, heart, liver, kidneys, bladder, spleen, mesenteric lymph nodes, abomasum, small and large intestine, epididymis, ovaries, fallopian tubes, uterus, thyroid gland, pancreas or adrenal glands. Comparative values for control animals were not available for this study (Cysewski et al., 1981).

The acute dermal toxicity of dicyclopentadiene 75 % (hydrocarbon mixture containing ca. 71 % endo-dicyclopentadiene) was studied in a limit test in 5 male and 5 female Sprague-Dawley CFY rats (210 to 256 g; 8 to 12 weeks old). The rats had 2000 mg/kg body weight applied to the intact dorsal skin for 24 h under occlusive conditions. No deaths were observed. Signs of systemic toxicity included hunched posture, lethargy, ruffled fur, red or brown staining of the muzzle and ptosis. Skin irritation with erythema and oedema was also observed. The rats were free of symptoms at the end of the 14-day observation period. Body weight gain was unimpaired. No gross effects were detected at autopsy. The dermal LD_{50} was therefore >2000 mg/kg body weight (Safepharm, 1989 a).

LD_{50} values reported for dermal toxicity in rabbits were 4.38, 4.99 and 6.60 g/kg body weight (24-hour exposure, occlusive cover). Observations on signs of systemic toxicity were not made by the authors (Kinkead et al., 1971; Smyth et al., 1954, 1962).

LC_{50} values determined after the inhalation of dicyclopentadiene are shown in Table 3. Based on these results, the substance is toxic.

All of the rats used (6/study) in acute inhalation studies died within 30 min (Smyth et al., 1962) or 2 h (Smyth et al., 1954) after administration of saturated dicyclopentadiene vapour (no further details provided). In a second study, the inhalation of 500 ppm (2700 mg/m³) dicyclopentadiene for 4 h led to the death of 1 of 6 rats, while 1000 ppm (5400 mg/m³) caused the deaths of all exposed rats (Smyth et al., 1962). In a further study by the same authors, 4 of

Table 3. Acute inhalation toxicity of dicyclopentadiene

Species	Number of animals/dose and sex	LC_{50} (mg/m³) (4-hour exposure)	Reference
Rat (Fischer 344)	6 male 6 female	1531 (6-hour exposure) 1906 (6-hour exposure)	Bushy Run, 1981
Rat	6 male	< 10,800	Smyth et al., 1954
Rat (Wistar)	2 male 2 female	< 5400	Gage, 1970
Rat	6 male 6 female	2700 – 5400	Smyth et al., 1962
Rat	6 male 6 female	1941 2080	Kinkead et al., 1971
Rat	–	610	Prokop'eva and Buzjun, 1988
Mouse (B6C3F1)	6 male 6 female	772 (6-hour exposure) 678 (6-hour exposure)	Bushy Run, 1981
Mouse	6 female	786	Kinkead et al., 1971
Mouse	-	400	Prokop'eva and Buzjun, 1988
Guinea-pig	6 male	4161	Kinkead et al., 1971
Rabbit	4 male	4163	Kinkead et al., 1971
Dog	1 female	2473–4174	Kinkead et al., 1971

-not specified

6 rats died within 14 days following inhalation of 2000 ppm (10,800 mg/m^3) for 4 h (Smyth et al., 1954). At a concentration of 2500 ppm (13,500 mg/m^3) inhaled for 1 hour, 1/4 rats died, while at 1000 ppm (5400 mg/m^3) inhaled for 4 h, all 4 exposed rats died (Gage, 1970).

The 4-hour LC_{50} value in a range-finding test was 660 ppm (3564 mg/m^3) in the rat. In more detailed studies by the same authors (determination of concentration by gas chromatography with a flame ionization detector) the 4-hour LC_{50} values fluctuated between, on the one hand, 145.5 (117.5 to 180.2) ppm (equivalent to 785.7 (634.5 to 973.1) mg/m^3) in the mouse, 359.4 (290 to 445.1) ppm (equivalent to 1940.8 (1566 to 2403.5) mg/m^3) in the male rat and 385.2 (311.1 to 477.1) ppm (equivalent to 2080 (1680 to 2576.3) mg/m^3) in the female rat and, on the other hand, 770.5 (595.6 to 996.6) ppm (equivalent to 4160.7 (3216.2 to 5381.6) mg/m^3) in the guinea pig and 771.0 (505.2 to 1177.0) ppm (equivalent to 4163 (2728.1 to 6355.8) mg/m^3) in the rabbit. The inhalation of 773 ppm (4174 mg/m^3) for 1 hour was fatal in a female dog. The inhalation of 458, 272 and ca. 68 ppm (equivalent to 2473, 1469 and 367 mg/m^3) for 4 h was survived by 1 female dog at each concentration. Dose-dependent eye irritation and disturbances of coordination were observed in all species; convulsions occurred when the animals died (Kinkead et al., 1971).

Groups of 6 male and 6 female F344 rats and B6C3F1 mice inhaled dicyclopentadiene at concentrations of 557, 260, 130 and 46 ppm (equivalent to 3009, 1404, 702 and 248 mg/m^3) for one 6-hour period. LC_{50} values of 283.6 ppm (1531 mg/m^3) for male rats and 352.9 ppm (1906 mg/m^3) for female rats were determined. In the mice, the LC_{50} values were 143 ppm (772 mg/m^3) in the males and 125.5 ppm (678 mg/m^3) in the females. Signs of toxicity observed included tonic-clonic spasms, loss of coordination, a reeling gait, bent backs, walking on the toes, stereotypical behaviour, irregular breathing, panting and a clear discharge from the nose (Bushy Run, 1981).

Subacute studies

No intolerance reactions were observed in beagle dogs following administration of 40, 125 or 375 ppm dicyclopentadiene in the diet for 14 days. The dose levels corresponded to ca. 5 to 12, 15 to 32 and 49 to 64 mg dicyclopentadiene/kg body weight/day. 1 dog was used per dose and sex. In this study, behaviour, feed consumption, body weight gain, haematology, clinical-chemistry and urine analyses, gross findings at autopsy and histology were used to evaluate the results (Litton Bionetics, 1976; Hart and Dacre, 1978). It should be noted, however, that even at the highest tested dose level, no impairment of body weight gain was recorded. A study carried out in accordance with current test guidelines would include a maximum tolerated dose (MTD).

In a study in 8-month-old mink, dicyclopentadiene was administered in the diet at doses of 1, 10, 100, 1000 and 10,000 ppm for 21 days. This was equivalent to about 0.17, 1.9, 18, 178 and 754 mg/kg body weight/day. 5 animals were used per dose and sex. The haematocrit value and differential blood count were not adversely affected at doses of up to 1000 ppm. There were no macroscopic or microscopic (brain, heart, lungs, kidneys, spleen and liver) changes compared with controls. In the first week of the study, weight gain was lower than that of the controls in both sexes, while in the second and third weeks of the study this effect was no longer seen. A mortality of 60 % was recorded at the highest dose level (10,000 ppm). At the same time there was a significant reduction in feed intake (about -75 % compared with controls) and body weight (about -66 % compared with controls) in this dose group. In the animals that were still alive at the end of the study (21 days) a significant reduction in the haematocrit value was observed as well as a lower number of neutrophilic granulocytes and a marked loss of body fat. The authors considered it possible that there was a connection between the reduced feed intake and the mortality and other observed side effects (Aulerich et al., 1979).

Groups of 6 male and 6 female F344 rats were treated with 0 (controls), 8, 40 or 200 mg dicyclopentadiene/kg body weight/day by gavage for 28 days. An additional 6 male and 6 female animals were observed for 14 days in the control and high dose groups. Body weight gain was inhibited in both sexes after administration of 200 mg/kg and in males after 40 mg/kg. At the end of the study, increased liver and adrenal gland weights and reduced thymus weights were seen in both sexes at the top dose level, and increased kidney weights were found in the males at the intermediate and high dose levels. Histologically, hypertrophy of the zona fasciculata of the adrenal gland and "foamy" cytoplasm in the hepatocytes was seen in both sexes at the top dose level. The histological effects were reversible after the 14-day observation period. Based on these findings, the no observed effect level was reported to be 8 mg/kg body weight/day (English abstract of a Japanese study; Satoh et al., 1990).

2 to 3 exposures to 300 mg dicyclopentadiene/m^3 led to aggression, altered behaviour, reduced chloride excretion, leukocytosis and anaemia in rats. Gross examination revealed haemorrhagic lungs (no further details; Prokop'eva and Buzjun, 1988).

In a subacute inhalation study in 2 male and 2 female Alderley Park rats per concentration, 1/4 rats died after inhaling 250 ppm (equivalent to 1350 mg/m^3) for 6 h/day 10 times; weight loss, nasal irritation, dyspnoea, lethargy, tremors and increased sensitivity were seen in the surviving animals. There were no pathological effects on the blood parameters studied (haemoglobin, erythrocytes, leukocytes, differential blood count, thrombocytes, clotting time, urea, sodium and potassium). The rats tolerated 15 6-hour exposures to 100 ppm (equivalent to 540 mg/m^3; Gage, 1970).

Groups of 6 male and 6 female Harlan-Wistar rats were exposed to 72, 146 and 332 ppm 98.3 % pure dicyclopentadiene vapour (equivalent to 389, 788 and 1793 mg/m^3) 7 h/day, 5 days/week over a total of 10 days. Compared with the controls, neither intolerance reactions nor macroscopically visible changes of the internal organs were observed after exposure to the two lower concentrations (72 and 146 ppm). At the high concentration (332 ppm), convulsions occurred in the males during the second day of inhalation and thereafter. All of the male and female rats died in this group after 1 to 4 days. Gross examination revealed that 332 ppm had caused bleeding in the lungs and gut in all cases and also in the thymus in the females (Kinkead et al., 1971).

The effect of dicyclopentadiene vapour (47, 72 and 146 ppm in the inspired air; equivalent to 254, 389 and 788 mg/m^3) on albino mice was studied with the same protocol and with the same number of animals as in the rat study described above. While the low dicyclopentadiene concentration (47 ppm) was tolerated without adverse reactions, exposure to 72 ppm led to the deaths of 5 of the 6 exposed mice within 10 days in both males and females. Body weight gain was normal in the 2 surviving mice. At the high dicyclopentadiene concentration (146 ppm), all of the animals died following convulsions during the first day of the study. No gross pathological effects were found during post-mortems of the mice that died during the study or those killed at the end of it (Kinkead et al., 1971).

In a further study, beagle dogs were exposed to dicyclopentadiene vapour in the air that they breathed at concentrations of 20, 47 and 72 ppm (equivalent to 108, 254 and 389 mg/m^3) in a protocol similar to that described above, but with 1 test animal per concentration. In this study, 20 ppm led to mild diarrhoea. This diarrhoea was more severe at 47 ppm, and increased salivation and defective coordination of the hind legs also occurred at this dose. At the high concentration (72 ppm) on the other hand, no such symptoms were detected. Body weight gain was normal in all animals. Macroscopic examination revealed no treatment-related effects (Kinkead et al., 1971).

7.3 Skin and mucous membrane effects

The uncovered application of 0.01 ml neat dicyclopentadiene to the clipped dorsal or abdominal skin of 5 albino rabbits caused moderate irritation of the skin (Smyth et al., 1954, 1962).

Following occlusive application of 2 g dicyclopentadiene/kg body weight to the clipped intact or scarified skin of rabbits (2 in each case) for 24 h, mild erythema was seen in 3 of the animals (no precise data provided). After 14 days, complete healing had occurred. No oedema, scab formation or systemic toxicity were observed (Litton Bionetics, 1976; Hart and Dacre, 1978).

0.5 ml Dicyclopentadiene 75 % (hydrocarbon mixture containing ca. 71 % endo-dicyclopentadiene) was applied to the clipped dorsal skin of 3 New Zealand white rabbits (2.2 to 2.5 kg) for 4 h under semi-occlusive cover. The skin findings were read 1, 24, 48 and 72 h after removal of the patch. The substance gave rise to well defined erythema and mild to severe oedema (primary irritation index 4.7). Signs of possible hyperkeratosis were observed on day 7. The substance was evaluated as moderately irritating (Safepharm, 1989 b).

The instillation of 0.5 ml neat dicyclopentadiene into the eye of the rabbit caused marked irritation of the mucous membranes (no further details; Smyth et al., 1954, 1962).

Eye irritation was tested in a further study in 9 New Zealand white rabbits according to the method of Draize. Irritation of the conjunctiva was seen in 7 of the exposed animals following instillation of 0.1 ml dicyclopentadiene (concentration unspecified). Even thorough rinsing of the eye 2 or 4 seconds after instillation of the substance did not give rise to any improvement. However, on the third day after instillation, the conjunctiva had returned to normal in all cases (Litton Bionetics, 1976; Hart and Dacre, 1978).

The eye irritancy of dicyclopentadiene 75 % (hydrocarbon mixture containing ca. 71 % endo-dicyclopentadiene) was tested in 3 New Zealand white rabbits (2.5 to 2.7 kg). The findings were evaluated 1, 24, 48 and 72 h after instillation of 0.1 ml into the right eye. 1 hour after instillation, clouding of the cornea was found in 2 animals, and inflammation of the iris and moderate irritation of the conjunctiva (which was still evident in 1 animal after 72 h) were seen in all animals. The substance was evaluated as moderately irritating (group score 18.3; Safepharm, 1989c).

7.4 Sensitization

Dicyclopentadiene was tested for its ability to induce sensitization in guinea pigs. 8 guinea pigs were given 10 intracutaneous injections (3/week) of 0.1 % dicyclopentadiene (weight/vol.). For the first injection, the volume administered was 0.05 ml, thereafter 0.1 ml was given. No skin reactions were observed. 2 weeks after the last dose, the animals were again injected with 0.05 ml of a 0.1 % solution of dicyclopentadiene as a challenge treatment. Only mild erythema was seen 24 and 48 h after administration of this challenge (number of animals reacting not specified, local reactions evaluated according to the Draize scheme). The result was evaluated as negative by the authors. 2,4-Dinitro-1-chlorobenzene, which was administered to 4 control guinea-pigs, caused a marked skin reaction (Litton Bionetics, 1976; Hart and Dacre, 1978).

The sensitizing potential of dicyclopentadiene 75 % (hydrocarbon mixture containing 71 % endo-dicyclopentadiene) was studied in a modified Buehler test in 12 albino (Dunkin-Hartley) guinea pigs, with 12 further animals serving as controls. During the induction phase, 0.5 ml of the neat substance was applied to the clipped skin (shoulder region) under occlusive cover for 6-hour periods on a total of 9 occasions within 3 weeks. On day 28, 0.2 ml of the substance was applied to the clipped skin of the right flank for 6 h under occlusive cover. The results were evaluated 24 and 48 h after removal of the patch. Under these study conditions, dicyclopentadiene 75 % was not sensitizing (Safepharm, 1989 d).

7.5 Subchronic and chronic toxicity

ARS-Sprague-Dawley rats received 80, 250 and 750 ppm dicyclopentadiene (98 to 99 % pure) in their diets for 90 days (equivalent to ca. 5 to 22, 16 to 62 and 44 to 198 mg dicyclopentadiene/kg body weight/day). 30 animals were used per dose and sex. No effects due to dicyclopentadiene treatment could be detected at any of the 3 tested doses. In a further study, dicyclopentadiene was administered to ICR-Swiss albino mice for 90 days at doses of 28, 91 and 273 ppm in the diet (equivalent to ca. 4 to 10, 12 to 31 and 37 to 88 mg/kg body weight/day). Each dose group consisted of 32 males and 32 females. No intolerance reactions were observed in this study either. The parameters studied in both experiments included behaviour, feed consumption and body weight gain, with haematological, clinical chemistry and urine tests being carried out, as well as post-mortems including the histology of 31 organs (Litton Bionetics, 1976; Hart and Dacre, 1978). It should be noted, however, that even at the highest tested dose level no inhibition of body weight gain was recorded relative to the controls. A study carried out in accordance with current test guidelines would include a maximum tolerated dose (MTD). Details of preliminary studies carried out to determine the doses administered were not provided.

Groups of 12 male and 12 female Harlan-Wistar rats were exposed to dicyclopentadiene (98.3 % pure) vapour in the air that they breathed at concentrations of 19.7, 35.2 and 73.8 ppm (equivalent to 106, 190 and 399 mg/m^3) 7 h/day, 5 days/week on 89 treatment days. Clinical examination for possible signs of toxicity and determinations of body weight were made on days 4, 13, 31, 55, 75 and 89 of the study. The liver and kidneys were weighed at the end of the study. After a comprehensive macroscopic examination, the heart, liver, kidneys, lungs, spleen, adrenal glands, trachea, prostate gland, testes, colon and mesenterium from each animal were examined microscopically. The absolute and relative kidney weights were significantly increased in the males at all 3 concentrations of dicyclopentadiene, although no concentration-dependency was evident. While no other effects were found at 19.7 ppm, body weight gain was impaired at the intermediate concentration (35.2 ppm). Microscopic examination showed effects on the kidneys, particularly in the males, including round cell infiltration, calcium and protein storage, dilation and degeneration of the kidney tubules and nephrolithiasis. At the high dicyclopentadiene concentration (73.8 ppm), these pathological effects were more severe. The body weight curve had clearly levelled off; on day 4 of treatment, the values were significantly below those in the controls. A significant increase in the relative liver weight was found in the males in this group, without histological effects being apparent. No treatment-related effects were found in the other microscopically-examined organs at any of the 3 dicyclopentadiene concentrations. The authors reported the no effect level as being between 19.7 and 35.2 ppm (Kinkead et al., 1971).

4 groups of 51 male and 51 female Fischer-344 rats and 45 male and 45 female B6C3F1 mice inhaled dicyclopentadiene vapour at analytical concentrations of 0 (control), 1.0, 5.1 or 51 ppm (equivalent to 5.4, 27.5, and 275 mg/m^3) 6 h/day, 5 days/week for 13 weeks (64 exposures in total). Groups of 9 animals were killed after 10, 30 and 64 exposures and after recovery phases of 29 and 92 days after the last exposure. Electron microscopy studies were carried out in an additional 3 rats/sex/group after 64 exposures and at the end of the observation period. None of the exposed rats died during the study, although 10 male and 9 female mice exposed to 51 ppm did. The apparent cause of death was congestion of the lungs accompanied, in a few cases, by kidney failure. No clinical signs of toxicity occurred in the rats or the mice. No effects on body weight gain were seen in the rats, while in the female mice in the highest dose group there was an increase in growth in the last week of the study, although this returned to normal during the recovery phase. An increase in absolute and relative liver weight was seen in

the male rats in the 51 ppm group after 10, 30 and 64 exposures. Urinalysis revealed a decrease in specific gravity and osmolality of the urine, which was dependent on the number of exposures and the dicyclopentadiene concentration. Increased urine volume was recorded in male animals exposed to 5.1 or 51 ppm. Changes in the excretion rate of calcium (decrease), sodium (decrease) and potassium (increase) occurred mainly in the male rats in the 51 ppm group. Creatinine clearance was unaffected. Epithelial cells were found in the urine sediment in male rats at all concentrations. Most of these effects were reversible at the end of exposure. Even after only 10 exposures, protein accumulation could be seen histologically in the epithelial cells of the proximal kidney tubules from the 5.1 and 51 ppm groups, although this was reversible during the recovery phase. Further histopathological findings in these groups included tubular hyperplasia and proteinosis, interstitial nephritis and thickening of the glomerular basal membrane. These effects partly or completely disappeared during the recovery phase, with the exception of the interstitial nephritis and thickening of the glomerular basal membrane. The authors were of the opinion that a possible explanation for the lesions, which did not clear up during the recovery period and which could also be seen in the control animals at this time point, was a so-called chronic glomerulonephrosis, which can occur spontaneously in male and female rats of this strain with increasing age (age of rats at the end of the study, 6 to 7 months). No similar effects on the kidneys were observed in the female rats or in the mice. The no effect level was given as 1 ppm by the authors (Bushy Run, 1982; Bevan et al., 1992).

Kinkead et al. (1971) determined a no effect level of between 19.7 and 35.2 ppm in rats following inhalation of dicyclopentadiene vapour. In contrast, Bevan et al. (1992) found a no effect level of 1 ppm. In the opinion of Bevan et al., this difference could be attributed to the proportion of cyclopentadiene monomers in the vapour produced. The dimer dicyclopentadiene is easily transformed into cyclopentadiene monomer at temperatures over 35 °C, a fact that was not known at the time of the study by Kinkead et al.. Bevan et al. therefore used a system that produced vapour below 35 °C in their study, in order to minimize the proportion of cyclopentadiene in the vapour produced. The dose-finding study for the 90-day study showed that dicyclopentadiene was considerably more toxic than cyclopentadiene on acute inhalation. The effects that occurred in the kidneys of the males were characterized by Bevan et al., based on more recent knowledge, as α2u-globulin nephropathy, which has no relevance for man (Borghoff, 1993). This knowledge was not available at the time that the Bushy Run (1982) study was carried out.

In a further study, groups of 3 male beagle dogs were exposed to air containing 8.9, 23.5 or 32.4 ppm dicyclopentadiene vapour (equivalent to 48, 127 and 175 mg/m^3) for 7 h/day, 5 days/week on 89 treatment days. Haematological and clinical chemistry analyses were undertaken 6 days before the beginning of the study and on days 20, 37, 65 and 85 of treatment (haematocrit, white blood count, blood urea, alanine aminotransferase, aspartate aminotransferase, acid and alkaline phosphatases). Tests on the urine were carried out 5 days before the beginning of the study and on days 21, 38, 68 and 87 of treatment. In addition, the animals were regularly observed for signs of toxicity. Liver and kidney weights were determined at the end of the study. No treatment-related effects occurred after exposure to 8.9 ppm. At the intermediate and high dicyclopentadiene concentrations (23.5 and 32.4 ppm) increases in the activities of acid and alkaline phosphatases and serum aspartate aminotransferase were determined on some study days, and an increased urea value and a lower number of neutrophilic granulocytes were also found. These findings, however, occurred only sporadically and were not concentration-dependent; they were considered by the authors to be without physiological significance. All other haematological and clinical-chemistry parameters studied remained unaffected. This was also true of the urinalysis. No pathological effects were evident on microscopic examination of the lungs, liver, kidneys, heart, spleen, adrenal glands, thyroid gland, parathyroid glands, oesophagus, diaphragm, lymph nodes, gall bladder, salivary glands, tongue, stomach, duodenum, pancreas, ileum, jejunum, colon, urinary bladder, prostate, testes, epididymides, brain, pituitary gland, skin or eyes. Because very slight pathological changes to the clinical chemistry parameters occurred on some study days at 23.5 ppm, the authors estimated the no effect level to be between 8.9 and 23.5 ppm (Kinkead et al., 1971).

7.6 Genotoxicity

7.6.1 In vitro

Dicyclopentadiene was tested for mutagenic activity in eukaryotes (*Saccharomyces cerevisiae* D3, test for mitotic recombination) and prokaryotes (*Salmonella typhimurium* TA 98, TA 100, TA 1535, TA 1537 and TA 1538). In *Saccharomyces cerevisiae* D3, concentrations ranged from 0.01 to 0.09 % in one test and from 0.01 to 0.017 % in a repeat test. In the Salmonella/microsome test, concentrations ranged from 10 to 5000 µg/plate in one test and from 10 to 500 µg/plate in the repeat test. The tests were repeated because toxic effects occurred at the higher dose levels. All of the tests were carried out with and without metabolic activation. The results of these studies, which were carried out in accordance with current test guidelines, gave no indications of any mutagenic activity (SRI, 1978).

In a further Salmonella/microsome test, dicyclopentadiene (95 % pure) was negative in the *Salmonella typhimurium* strains TA 98, TA 100, TA 1535 and TA 1537 at concentrations ranging from 0 to 333 µg/plate, with and without metabolic activation (S9-mix from Aroclor 1254-induced rat and hamster liver). The highest concentration was cytotoxic (Zeiger et al., 1987).

7.6.2 In vivo

No information available.

7.7 Carcinogenicity

The question of the possible carcinogenic properties of dicyclopentadiene was investigated by the Institute of Chemical Biology at San Francisco University in 1975, on behalf of the National Cancer Institute. After intramuscular administration of dicyclopentadiene to rats under the chosen study conditions (which were not further described), there were no indications that dicyclopentadiene was carcinogenic (no further details; Rosenblatt et al., 1975).

A carcinogenicity study on dicyclopentadiene is planned as part of the National Toxicology Program (NTP; NTP, 1992).

7.8 Reproductive toxicity

Groups of 20 pregnant Sprague-Dawley rats were given dicyclopentadiene in their diet at concentrations of 0, 80, 250 and 750 mg/kg feed (ppm) from day 6 to day 15 of pregnancy. The 750 ppm concentration was equivalent to a substance intake of 150 mg/kg body weight at an assumed feed consumption of 200 g/kg body weight/day. The numbers of implantations, live and dead foetuses and resorptions were recorded on day 19. No treatment-related effects or changes in reproductive ability were seen in the dams. The foetuses showed no skeletal or visceral malformations and there was no change in the sex ratio (Litton Bionetics, 1980).

3-month-old mink were given dicyclopentadiene at doses of 100, 200, 400 and 800 ppm (equivalent to 23.6, 42.4, 85.0 and 169.9 mg/kg body weight/day) for 12 months in the diet. Groups of 3 females and 1 male were kept together in a cage, and 4 such groups were used per dose. The study was arranged so that mating, birth and rearing of the young took place in the last third of the study. The following study parameters were assessed: mortality, body weight gain (determined every other week), feed intake (determined every other week), analysis of the blood (haematocrit and differential blood count every 3 months), duration of pregnancy, litter size, weight of offspring, sex ratio of the offspring and mortality. The following organs from the parent animals were weighed at the end of the study: brain, liver, kidneys, spleen, gonads, lungs, heart and adrenal glands. An additional 5 organs were fixed for histological examination. There were no adverse effects on mortality, body weight gain, feed consumption, blood count or reproductive behaviour (fertility of the males, duration of pregnancy, litter size, litter weight and sex ratio of the offspring). The offspring in the 200, 400 and 800 ppm groups showed significantly reduced body weights 4 weeks after withdrawal of the substance. Mortality in the offspring was not significantly affected up to this time point. Macroscopic and microscopic examination of the organs at the end of the study gave no reliable indications of treatment-related effects. However, spleen weights were significantly reduced at 400 ppm and testes weights at 800 ppm, although there were no pathological effects on histological examination of the testes. The authors considered that these were probably chance findings (Aulerich et al., 1979).

In a 3-generation study, groups of 10 male and 10 female rats were given a diet containing dicyclopentadiene at concentrations of 0, 69.3 or 693 ppm. The highest concentration was equivalent to a daily intake of 50 mg of the substance/kg body weight at an assumed feed intake of 70 g/kg body weight/day. During pregnancy, a substance intake of 140 mg/kg body weight was calculated at an assumed feed intake of 200 g/kg body weight/day. No treatment-related effects were found after studying the litters, fertility-, survival- and lactation indices, body weights and feed intake, or examining the parent, F_1, F_2 and F_3 generations macroscopically (Litton Bionetics, 1980).

7.9 Effects on the immune system

No information available.

7.10 Neurotoxicity

Groups of 10 male and 10 female F344 rats (6 to 7 weeks old) inhaled dicyclopentadiene vapour for 6 h daily for 9 days at analytical concentrations of 99.9, 33.0 or 5.1 ppm (equivalent to 539, 178 and 28 mg/m^3). A modified Irwin screen was carried out before and after exposure, on days 1, 2, 5, 6 and 7, and before autopsy. At the highest concentration, body weight gain in the males and females was significantly lower than that in the controls. In addition, a few animals showed a red-black encrusted nasal discharge. The relative kidney weights in the male rats were significantly higher than in the controls at all three concentrations. The Irwin screen revealed an abnormal righting reflex following the first exposure in one male at the highest concentration. No further treatment-related effects were established (Bushy Run, 1981).

Groups of 8 to 9 male and 10 female B6C3F1 mice were also exposed in the study protocol described above. Stereotypical behaviour consisting of constant repetition of a pointless body movement was observed in both sexes at the highest concentration. All mice exposed to 99.9 ppm died by the end of the fifth exposure. At this concentration, reduced motor activity, abnormal tail and toe reflexes following pinching, abnormal gait, difficulties in coordination and slight trembling were also observed as well as, in the females, abnormal righting reflexes and convulsions. 2 males and 4 females in the 33 ppm group and 1 male in the 5 ppm group showed a reduced response to irritant situations and stimuli. These effects, however, were only transitory and were also seen in 2 female control mice and were thus not statistically significant. Alopecia and abnormal tail and toe reflexes following pinching were observed to the same extent at the two lower concentrations and in the controls. Body weight gain was not impaired and no treatment-related lesions were found (Bushy Run, 1981). These studies gave no indications that the substance had neurotoxic properties.

7.11 Other effects

In order to study the possible effect of dicyclopentadiene on the hexobarbital-sleeping time in rats (Charles River-COBS-CD (SD)BR),1 group containing 10 rats of each sex were given a diet containing 750 ppm dicyclopentadiene. A group of 10 rats of each sex served as control. On day 5 of dicyclopentadiene administration, all rats were given 100 mg hexobarbital/kg body weight by intraperitoneal injection. No change in sleeping time occurred compared with the controls. (Litton Bionetics, 1976; Hart and Dacre, 1978). Based on this result, no enzyme-inducing effect of dicyclopentadiene can be assumed.

8. Experience in humans

2 volunteers were exposed to 1 ppm dicyclopentadiene (equivalent to 5.4 mg/m^3) for 30 min in an inhalation chamber. One subject experienced mild eye and throat irritation after 7 min while the other reported a diminished sense of smell. After a 30-minute break, the same 2 subjects were exposed to a dicyclopentadiene concentration of 5.5 ppm for another 30 min; neither reported any effects on their sense of smell but one experienced eye irritation after 10 min and the second could still taste dicyclopentadiene for 1 h after the end of exposure (Kinkead et al., 1971).

During a 5-month period of inadvertent exposure to dicyclopentadiene, it was established that occasional temporary headaches occurred during the first 2 months. These reactions did not occur during the following 3 months. The authors interpreted this as a sign of a slowly increasing tolerance level (Kinkead et al., 1971).

3 gamekeepers who investigated a pond in the Rocky Mountain Arsenal reported severe headaches and irritation of the skin and throat following contact with sick or dead ducks that were exposed to contaminants in the pond. Dicyclopentadiene was subsequently identified as a contaminant in the pond and was considered to be responsible for the symptoms that occurred (NIOSH, 1981).

Irritation of the mucous membranes is expected to occur in people exposed to dicyclopentadiene concentrations of 10 ppm (equivalent to 54 mg/m^3) and above (no further details; ACGIH, 1980).

The olfactory threshold for dicyclopentadiene was reported to be 0.003 ppm in man (Kinkead et al., 1971). However, the value was based on a test in only 3 subjects. These were volunteers and their sense of smell was not checked prior to the experiment.

9. Threshold limit values

The TLV value in the USA is 5 ml/m^3 (ppm; equivalent to 27 mg/m^3 air; ACGIH, 1993/94).

The MAK value for dicyclopentadiene in the Federal Republic of Germany is set at 0.5 ml/m^3 (ppm; equivalent to 3 mg/m^3; DFG, 1993; TRGS, 1994).

References

ACGIH (American Conference of Governmental Industrial Hygienists Inc.)
Documentation of the threshold limit values
4th ed., p. 137–138 (1980)

ACGIH (American Conference of Governmental Industrial Hygienists Inc.)
Threshold limit values for chemical substances and physical agents and biological exposure indices 1993–1994

Aulerich, R.J., Coleman, T.H., Polin, D., Ringer, R.K., Howell, K.S., Jones, R.E., Kavanagh, T.J.
Toxicology study of diisopropyl methylphosphonate and dicyclopentadiene in mallard ducks, bobwhite quail, and mink
Poultry Science Department, Michigan State University, East Lansing, Michigan, USA, Report AD-A087 257 (1979)

Bevan, C., Snellings, W.M., Dod, D.E., Egan, G.F.
Subchronic toxicity study of dicyclopentadiene vapor in rats
Toxicol. Ind. Health, 8 (6), 353–367 (1992)

Borghoff, S.J.
α2u-Globulin-mediated male rat nephropathy and kidney cancer: relevance to human risk assessment
CIIT Activities, 13 (4), 1–8 (1993)

Bushy Run Research Center
Dicyclopentadiene - 9-day rat and mouse inhalation study
Unpublished report no. 43–527 (1981)
On behalf of the EXXON Corporation
See also NTIS/OTS0535718

Bushy Run Research Center
Dicyclopentadiene – ninety-day vapor inhalation study on rats and mice
Unpublished report no. 44–520 (1982)
On behalf of the EXXON Corporation

Cysewski, S.J., Palmer, J.S., Crookshank, H.R., Steel, E.G.
Toxicological evaluation of diispropyl methylphosphonate and dicyclopentadiene in cattle
Arch. Environ. Contam. Toxicol., 10, 605–615 (1981)
and NTIS/AD-A093 673

DFG (Deutsche Forschungsgemeinschaft), Senatskommission zur Prüfung gesundheitsschädlicher Arbeitsstoffe
MAK- und BAT-Werte-Liste 1993
VCH Verlagsgesellschaft, Weinheim (1993)

Dow Chemical
AIDA-Grunddatensatz 4,7-Methano-1H-indene,3a,4,7,7a-tetrahydro-9CI (1992)

Gage, J.C.
The subacute inhalation toxicity of 109 industrial chemicals
Br. J. Ind. Med., 27, 1–18 (1970)

Griesbaum, K., Hönicke, D., Olson, M.
Cyclopentadiene and cylopentene
In: Ullmann's encyclopedia of industrial chemistry
5th ed., vol. A8, 227–235
VCH Verlagsgesellschaft, Weinheim (1987)

Hart, E.R., Dacre, J.C.
Mammalian toxicologic studies on dicyclopentadiene
In: Plaa, G.L., Duncan, W.A.M. (eds.)
Proceedings of the 1st international congress on toxicology as a predictive science, p. 448–449 (1978)

Ivie, G.W., Oehler, D.D.
Fate of dicyclopentadiene in a lactating cow
Bull. Environ. Contam. Toxicol., 24, 662–670 (1980)

Kinkead, E.R., Pozzani, U.C., Geary, D.L., Carpenter, C.P.
The mammalian toxicity of dicyclopentadiene
Toxicol. Appl. Pharmacol., 20, 552–561 (1971)

Litton Bionetics Inc.
Mammalian toxicological evaluation of DIMP and DCPD
Report contract-no. DAMD 17–75-C-5068 (1976)
On behalf of the U.S. Army Medical Research and Development Command, Washington D.C.
NTIS/AD-A058 323

Litton Bionetics Inc.
Mammalian toxicological evaluation of DIMP and DCPD (phase 2)
Report on contract no. DAMD 17–77-C-7003 (1980)
On behalf of the U.S. Army Medical Research and Development

Command, Washington D.C.
Cited in: JACC report no. 19
Dicyclopentadiene
ECETOC (European Chemical Industry Ecology and Toxicology Centre), Brussels (1991)

NIOSH (National Institute for Occupational Safety and Health)
Health Hazard Evaluation Report HETA 81–176–968, Rocky Mountain Arsenal, Basin F, Commerce City, Colorado
NTIS (National Technical Information Service) (1981)

NTP (National Toxicology Program)
Executive summary of safety and toxicity information - dicyclopentadiene
Executive Committee Draft Report (1990)

NTP (National Toxicology Program)
Review of current DHHS, DOE, and EPA research related to toxicology
Fiscal year 1992

Prokop'eva, A.S., Buzjun, A.F.
Neue Daten zur Toxizität von Dicyclopentadien (German translation of the Russian)
Gig. Tr. Prof. Zabol., 8, 57 (1988)

Rosenblatt, D.H., Miller, T.A., Dacre, J.C., Muul, I., Cogley, D.R.
Problem definition studies on potential environmental pollutants. II. Physical, chemical, toxicological, and biological properties of 16 substances
U.S. Army Medical Bioengineering Research and Development Command, Technical Report 7509, J1-J8 (1975)
NTIS/AD-A030 428

RTECS (Registry of Toxic Effects of Chemical Substances)
4,7-Methanoindene,3a,4,7,7a-tetrahydro, RTECS no. PC1050000
produced by NIOSH (National Institute for Occupational Safety and Health) (1992)

Safepharm Laboratories Ltd.
DCPD 75 %: acute dermal toxicity (limit test) in the rat
Unpublished report no. DET 1227 (1989 a)
On behalf of Dow Europe

Safepharm Laboratories Ltd.
DCPD 75 %: acute dermal irritation test in the rabbit
Unpublished report no. DET 1228 (1989 b)
On behalf of Dow Europe

Safepharm Laboratories Ltd.
DCPD 75 %: acute eye irritation test in the rabbit
Unpublished report no. DET 1229 (1989 c)
On behalf of Dow Europe

Safepharm Laboratories Ltd.
DCPD 75 %: modified nine-induction Buehler contact sensitization study in the guinea pig
Unpublished report no. DET 1230 (1989 d)
On behalf of Dow Europe

Safepharm Laboratories Ltd.
DCPD 75 %: acute oral toxicity test in the rat
Unpublished report no. DET 1232 (1989 e)
On behalf of Dow Europe
cited in: JACC Report no. 19
Dicyclopentadiene
ECETOC (European Chemical Industry Ecology and Toxicology Centre), Brussels (1991)

Satoh, M., Okamiya, H., Furukawa, F., Shinoda, K., Imazawa, T., Toyada, K., Takahashi, M.
Twenty-eight day repeated dose toxicity test of dicyclopentadiene in F344 rat (English abstract of a Japanese report)
Eisei Shikensho Hokoku, 108, 71–77 (1990)

Sax, N.I., Lewis, R.J. (eds.)
Rapid guide to hazardous chemicals in the workplace, p. 59
Van Nostrand Reinhold Company, New York (1986)

Scherb, H.
Cyclopenten und -pentadien
In: Ullmanns Encyklopädie der technischen Chemie
4th ed., vol. 9, p. 699–704
Verlag Chemie, Weinheim (1975)

Smyth, H.F., Jr., Carpenter, C.P., Weil, C.S., Pozzani, U.C.
Range-finding toxicity data: list V
Am. Ind. Occup. Med., 10, 61–68 (1954)

Smyth, H.F., jr., Carpenter, C.P., Weil, C.S., Pozzani, U.C., Striegel, J.A.
Range-finding toxicity data: list VI
Arch. Ind. Hyg. Assoc. J., 23, 95–107 (1962)

SRI International -
In vitro microbiological mutagenicity assays of dicyclopentadiene
Report on SRI project LSU-5612 (1978)
Prepared for Environmental Protection Agency, National Environmental Research Center, Water Supply Research Laboratory, Cincinnati, Ohio

TRGS (Technische Regeln für Gefahrstoffe) 900
Neufassung der TRGS 900
Bekanntmachung des Bundesministers für Arbeit und Sozialordnung
Bundesarbeitsblatt 6/1994, pp 34–53

Zeiger, E., Anderson, B., Haworth, S., Lawlor, T., Mortelmans, K., Speck, W.
Salmonella mutagenicity tests: III. Results from the testing of 255 chemicals
Environ. Mutagen., 9, Suppl. 9, 1–110 (1987)

Acetic acid methyl ester

1. Summary and assessment

Acetic acid methyl ester can be absorbed through the lungs and also through the gastro-intestinal tract and the skin. The substance is rapidly hydrolysed to methanol and acetic acid by the non-specific esterases present in the blood and tissues. Thus, following oral administration to rabbits only methanol, and not the unmetabolized ester, has been detected in the blood and urine. In addition, comparative exposure studies with acetic acid methyl ester in volunteers have shown that its metabolites, methanol and acetic acid, are responsible for the toxic profile of the substance.

The acute toxicity of acetic acid methyl ester on oral and dermal exposure is low (LD_{50} rat oral > 5000 mg/kg body weight, LD_{50} rabbit oral 3700 mg/kg body weight; LD_{50} rat dermal > 2000 mg/kg body weight, LD_{50} rabbit dermal > 5000 mg/kg body weight). The inhalation of high concentrations of the substance causes eye irritation, dyspnoea, narcosis and deaths in rats, mice and cats. In cats exposed by inhalation for 6 h/day for 8 days at 19 to 21 mg acetic acid methyl ester/l, the local irritant effects were prominent.

Acetic acid methyl ester is not irritating to the rabbit skin, but is irritating to the rabbit eye.

Acetic acid methyl ester does not induce gene mutations in bacterial systems in the Ames test or in eukaryotic systems in *Saccharomyces cerevisiae* D61.M. It has a pronounced effect on the induction of gene conversion by cyclophosphamide in the *Saccharomyces cerevisiae* D_4 test system (marked reduction in the frequency of gene conversion). However, it is not possible to draw any conclusions on the mutagenic potential of the substance alone from this combined effect. Acetic acid methyl ester induces unequivocal mitotic chromosomal malsegregation, that is genome mutation, in *Saccharomyces cerevisiae* test system D61.M.

In a chronic inhalation study in rats with a varnish thinner, which contained acetic acid methyl ester (12.6%) as a component in addition to toluene (52.8%) and isopropanol (21.7%), adaptive and degenerative changes in the cerebral cortex have been observed by electron microscopy.

In man, acetic acid methyl ester is neither a skin irritant nor a sensitizer, but exposure to the vapour for 5 min leads to irritation of the eyes, nose, throat and trachea from ca. 4950 ppm.

No reports of industrial experience or cases of poisoning with pure acetic acid methyl ester in man are available. The ester commonly occurs as a component of solvent mixtures. Repeated inhalation exposure to pigment thinners which, in addition to methanol, contain acetic acid methyl ester as a main component have, in two cases, led to disturbances in vision and blindness. In solvent-exposed workers in the paint industry (who were exposed to acetic acid methyl ester among other substances), impaired CNS function (reaction times, speed of perception, short-term memory) has been observed, but there were no hepatotoxic effects (liver enzyme activities in serum not increased) or genotoxic effects (sister chromatid exchange rate and chromosome aberration rate in lymphocytes not increased). Besides toluene and the ethyl and butyl esters of acetic acid, acetic acid methyl ester (a component of the thinning solution in the polish) is also regarded as responsible for the occurrence of corneal changes in the eyes of furniture polishers (polishers' keratitis).

The olfactory threshold for acetic acid methyl ester in the air is between 4.6 and 300 ppm.

The MAK value in the Federal Republic of Germany is 200 ppm (or 610 mg/m^3), the TLV value in the USA is 200 ppm (or 606 mg/m^3).

2. Name of substance

2.1	Usual name	Acetic acid methyl ester
2.2	IUPAC-name	Acetic acid methyl ester
2.3	CAS-No.	79-20-9
2.4	EINECS-No.	201-185-2

3. Synonyms, common and trade names

Devoton
Essigsäuremethylester
Methyl acetate
Methyl ethanoate
Tereton

4. Structural and molecular formulae

4.1 Structural formula

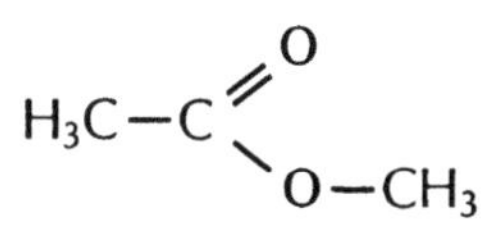

4.2 Molecular formula $C_3H_6O_2$

5. Physical and chemical properties

5.1 Molecular mass, g/mol 74.08

5.2 Melting point, °C
–99 (Verschueren, 1983)
–98.1 (Weast, 1988)
–98 (Budavari et al., 1989; Falbe and Regitz, 1990; Hoechst, 1994 a; von Oettingen, 1960)

5.3 Boiling point, °C
56–58 (DIN 53171) (Hoechst, 1994 b)
56.9 (Budavari et al., 1989)
57 (Weast, 1988)
57.3 (von Oettingen, 1960)
58 (Falbe and Regitz, 1990)

5.4 Vapour pressure, hPa
217 (at 20 °C) (Hoechst, 1994 a)
220 (at 20 °C) (DFG, 1994)
227 (at 20 °C) (Verschueren, 1983; von Oettingen, 1960)

		313 (at 25 °C) (Verschueren, 1983) 340 (at 30 °C) (Verschueren, 1983)
5.5	Density, g/cm³	0.928 (at 20 °C) (Falbe and Regitz, 1990; Flury and Wirth, 1933) 0.93 (at 20 °C; DIN 51757) (Hoechst, 1994 a) 0.9330 (at 20 °C) (Weast, 1988) 0.9342 (at 20 °C) (Budavari et al., 1989) 0.9279 (at 25 °C) (Budavari et al., 1989) 0.93 (at 25 °C) (Verschueren, 1983)
5.6	Solubility in water	240 g/l (at 20 °C) (Verschueren, 1983) 250 g/l (at 20 °C) (Hoechst, 1994 b)
5.7	Solubility in organic solvents	Soluble in ethanol, diethyl ether, acetone, benzene and chloroform (Budavari et al., 1989; Weast, 1988)
5.8	Solubility in fat	n-Octanol/water partition coefficient, log Pow: ca. 0.18 (measured) (Leo et al., 1971) 0.35 (calculated) (Leahy, 1986)
5.9	pH value	Neutral (Hoechst, 1994 b)
5.10	Conversion factor	1 ml/m³ (ppm) ≙ 3.02 mg/m³ 1 mg/m³ ≙ 0.33 ml/m³ (ppm) (at 1013 hPa and 25 °C)

6. Uses

Solvent for nitrocellulose, acetylcellulose, esters and ethers (Falbe and Regitz, 1990; Aguilo et al., 1985).

7. Experimental results

7.1 Toxicokinetics and metabolism

It has been reported that acetic acid methyl ester is absorbed through the lungs and the gastro-intestinal tract (no further details; von Oettingen, 1960).

Following inhalation of acetic acid methyl ester by cats (5 animals) at concentrations of 60 to 120 mg/l for 20 to 140 min and by rabbits (2 animals) at concentrations of 60 to 250 mg/l for 153 to 155 min, a reduction in the pH-value of the blood by 0.11 to 1.04 was measured (Flury and Wirth, 1933).

The mechanism of toxicity and the metabolism of various acetic acid esters (the methyl, ethyl, isopropyl, butyl and isoamyl esters) was studied in rabbits. In each case, 20 ml/kg body weight of a 5% aqueous solution of the ester was administered by gavage. Blood, urine and gastric juice samples were taken from the animals before administration, 30 min after administration and then at hourly intervals for up to 5 h, prior to analysis for their ester and alcohol content by means of gas chromatography. The administered acetic acid esters could not be detected in any of the samples, but their corresponding alcohols were detectable in the blood and urine after a short time. Only 30 min after gavaging with acetic acid methyl ester, a relatively high concentration of methanol was found in the blood, which continued to increase slightly for up to 3 h after administration and then fell somewhat. After administration of the other acetic acid esters, the concentrations of the corresponding alcohols in the blood were lower than the methanol concentration after administration of acetic acid methyl ester (see Table 1).

Table 1. Blood alcohol concentrations following oral administration of various acetic acid esters to rabbits

Alcohol of the acetic acid ester	Concentration (mg/ml) after 30 min	1 h	2 h	3 h	4 h	5 h
Methanol	0.446	0.513	0.456	0.573	0.416	0.345
Ethanol	0.373	0.341	0.244	0.163	0.077	–
Isopropanol	0.083	0.110	0.081	0.067	0.050	0.038
Butanol	0.047	0.061	0.019	0.006	–	–

The concentration of methanol in the urine was also measured for up to 4 h after administration of acetic acid methyl ester. Relatively large amounts were already present half an hour after administration of the substance, the highest concentrations were excreted after 2 h, but the methanol concentration in the urine was still high after 4 h. The concentrations of the alcohols of the other acetic acid esters were clearly lower, but increased, in most cases continuously, during the study period (see Table 2).

Table 2. Alcohol concentrations in the urine following oral administration of various acetic acid esters to rabbits

Alcohol of the acetic acid ester	Concentration (mg/ml) after 30 min	1 h	2 h	3 h	4 h	5 h
Methanol	0.133	0.294	0.480	0.365	0.450	–
Ethanol	–	0.060	0.071	0.125	0.140	–
Isopropanol	0.037	0.102	0.102	0.084	0.064	0.043
Butanol	0.015	0.030	0.025	0.010	0.010	0.010

The stomach contents were also analyzed for acetic acid esters and alcohols at the above-mentioned time intervals. According to the author, the results were similar for all 4 esters studied, but these were only presented in detail for acetic acid ethyl ester. The hydrolysis of acetic acid ethyl ester to ethanol and acetic acid took place slowly in the stomach, and 2 hours after administration 0.8 mg acetic acid ethyl ester/ml were still detectable. At this time point the original compound could no longer be detected in the blood and urine and only ethanol could be detected analytically. This shows that acetic acid ethyl ester is partially hydrolysed in the stomach. The main hydrolysis occurs during absorption, by the esterases contained in the gastric mucosa and the esterases in the blood itself. This explains why only the alcohols and acetic acid and not the original substance were detectable in the blood and urine. The author indicated that hydrolysis also occurred after inhalation exposure, and only the hydrolysis products were present in the blood and urine. Corresponding results were also given for acetic acid methyl ester. Details of the studies on the gastric juice were not, however, provided in the report. At the same time intervals, the pH-value of the blood of the animals was determined with the aid of a blood gas analyzer (no further details; see Table 3).

Table 3. pH-values of the blood following oral administration of various acetic acid esters to rabbits

Acetic acid ester	pH-values before administration	pH-values after				
		30 min	1 h	2 h	3 h	4 h
Methyl acetate	7.330	7.140	7.220	7.235	7.270	7.210
Ethyl acetate	7.391	7.022	7.170	7.290	7.376	7.410
Isopropyl acetate	7.373	7.061	7.105	7.276	7.250	7.253
Butyl acetate	7.355	7.166	7.198	7.210	7.251	7.313
Isoamyl acetate	7.372	7.182	7.252	7.294	7.395	7.410

All 5 acetic acid esters gave rise to acidosis between 30 and 60 min after administration, which declined over the course of 2 to 3 h. The most marked acidosis was measured after administration of the ethyl and isopropyl esters of acetic acid. The acidosis after administration of the methyl and butyl esters and particularly after administration of the amyl ester was less pronounced. The author considered that the acidosis was caused by the acetic acid arising during hydrolysis. There was also an increase in the partial pressure of CO_2 and a decrease in the partial pressure of oxygen after administration of the various acetic acid esters, effects which were probably a consequence of the acidosis and should be regarded as an indication of a reduction in the speed of metabolism of the acetic acid esters (Tambo, 1973).

In order to study the rate of cleavage of various esters of acetic acid (methyl, ethyl, butyl and amyl) in vitro in rat and human blood, 8 parts of ester solution (in physiological saline) were mixed with 5 parts of blood in a 25 ml glass, and then kept at 37 °C in a thermostat. At periodic intervals, 1 ml samples of the solution were removed for quantitative determination of the esters in accordance with the method of Chestrin. For each substance and type of blood, 3 to 4 studies were carried out. The esters were cleaved under these in vitro conditions. According to the author, the reduction of the ester concentrations per unit time resembled a first-order process in form, which agrees with the concept of the biological hydrolysis of esters occurring in the manner of a first-order reaction. The acetic acid esters were cleaved more rapidly in the blood from albino rats than in human blood. In the albino rat blood the rate of cleavage of the esters increased with increasing chain length from the methyl to the amyl ester. In human blood the cleavage rates of the methyl and ethyl esters were identical, while for the butyl and amyl esters, ca. 10-fold and ca. 6–7-fold higher cleavage rates, respectively, were found. The author assumed that acetic acid esters are also cleaved in the organism in the same way and probably even more quickly than in vitro (Filov, 1961).

The same investigator studied the absorption of the methyl, butyl and amyl esters of acetic acid through the lungs in tracheotomized rabbits, which inhaled the ester vapours from a 645-l chamber. Blood samples were taken periodically to determine the ester level. In addition, the ester concentrations were determined in the inhaled and exhaled air at frequent intervals during the studies which lasted for up to 3 h. As the studies with the various esters revealed a consistent picture, only the results obtained with acetic acid ethyl ester were presented in detail as an example in the report. The esters were detectable in the exhaled air within a short period of the beginning of inhalation and equilibrium was established within the first minutes. The ester concentrations in the exhaled air were 30 to 50% of the concentrations measured in the inhaled air (up to 20 mg/l). At the end of the inhalation exposure the esters disappeared from the exhaled air. Despite the observed retention of the esters during inhalation exposure, they were not detected in the blood in any of the studies, a finding which is indicative of their rapid cleavage in the body. The author considered that the slower cleavage of the methyl and ethyl esters of acetic acid in vitro compared with that in vivo, was of less significance when the quantitative relationship between the blood and the esters was considered; in the in vitro studies large amounts of ester were added to the blood, while in vivo, a gradual diffusion of the ester into the continually renewed blood takes place. Based on these results, the author concluded that the toxicity of the esters studied is determined to a large extent by the properties of their cleavage products (Filov, 1961).

In a further study, rats were placed in 25-l bottles in which high, possibly saturated, concentrations of the compounds were achieved by the application of the methyl or amyl ester of acetic acid to filter paper. After 10 to 20 min the rats became narcotic (lying on the side, absence of conscious muscle movements). In the two studies with acetic acid methyl ester, 70 and 80 mg% were found in the blood at this time following decapitation. Acetic acid amyl ester, which is cleaved more rapidly than the methyl ester, was not detectable in the blood of the rats under these study conditions (Filov, 1961).

In order to study the question of whether the narcotic effects of acetic acid methyl ester and acetic acid ethyl ester are caused by the esters themselves or by the corresponding alcohols which are rapidly liberated in the body, albino rats were placed in glass bottles (volume 25 l) with high concentrations of the esters. After the onset of narcosis, characterized by lying on the side and the absence of spontaneous muscle movements (after ca. 10 to 20 min) the animals were decapitated and their blood was analyzed for its ester and methanol or ethanol content. For each ester, 4 studies were carried out. Analogous studies were undertaken with methanol and ethanol vapour. The concentrations of the esters in the blood were determined by the method of Hestrin. For methanol, the method of Feldstein and Klendshoj was used, while for ethanol, that of Widmark was employed. As these methods of determining alcohols do not permit any differentiation between alcohols and esters, the actual alcohol concentrations were derived from the analytically determined methanol and ethanol concentrations by subtracting the values determined for the methyl and ethyl esters of acetic acid.

Table 4. Concentration of esters and alcohols in the blood of rats after inhalation of acetic acid methyl ester and ethyl ester vapours and the methanol and ethanol content after inhalation of these alcohols

Substance (vapour)	Concentration (in mg%) Ester	Alcohol	Methanol	Ethanol
Methyl acetate	72 61 72 102	54 19 38 25		
Ethyl acetate	28 24 27 21	90 66 132 46		
Methanol			465 394 387 400	
Ethanol				380 416 415 416

At the point at which the animals were seen to be lying on their sides, the blood concentration of acetic acid methyl ester was found to be higher than that of acetic acid ethyl ester. The methanol level in the blood of the rats at this time, following inhalation of the vapours of this alcohol, was on average 10 times higher than on inhalation of acetic acid methyl ester. Under the same conditions, the ethanol content of the blood was on average 6 times higher following inhalation of the alcohol than after inhalation of the corresponding ester. Thus the rat studies showed that, on inhalation of high concentrations of acetic acid methyl and ethyl esters, the narcotic effect was produced preferentially by the esters themselves and to a lesser extent by the alcohols arising from them (Filov, 1964 a, b).

Most acetic acid esters tested (the n-butyl, sec.-butyl, tert.-butyl, ethyl, amyl, methyl and n-propyl esters), have been shown to be unstable in the blood in vitro and are rapidly hydrolysed to the corresponding alcohol and acid. Because of this rapid hydrolysis, the solubility coefficients (lambda and alpha) of the acetic acid esters in the blood could not be measured. Because of the instability of the acetic acid esters the authors recommended that in monitoring studies involving acetic acid ester-exposed subjects, the corresponding alcohol rather than the acetic acid ester should be measured in the breath (Ghittori et al., 1984).

7.2 Acute and subacute toxicity

The acute oral LD_{50} for acetic acid methyl ester in the rat was given as > 5000 mg/kg body weight (no further details; Moreno, 1976).

Following oral administration of acetic acid methyl ester to Carworth-Wistar rats (4 to 5 weeks old, weight 90 to 120 g), the LD_{50} value was 6.97 ml/kg body weight (equivalent to ca. 6482 mg/kg body weight). The observation period was 14 days (Smyth et al., 1962).

In rabbits, the ND_{50} (narcotic dose) after oral administration was 3260 mg acetic acid methyl ester/kg body weight and the LD_{50} was 3700 mg/kg body weight. Signs of toxicity included impaired movement, loss of corneal reflexes, nystagmus, dyspnoea and bradycardia (Munch, 1972).

The oral administration of 1500 mg acetic acid methyl ester/kg body weight to rabbits was survived without pathological effects (no further details; Flury and Wirth, 1933).

A study of the acute dermal toxicity of acetic acid methyl ester (99.9% pure) in a limit test in Wistar rats (weight ca. 200 g) in accordance with OECD guideline no. 402 gave an LD_{50} value of > 2000 mg/kg body weight for male and female rats. Neither deaths nor clinical signs of toxicity occurred following administration of 2000 mg/kg body weight. Body weight gain was normal. There were no macroscopic effects on the animals examined after the 14-day observation period (Hoechst, 1988 a).

The acute dermal LD_{50} in rabbits was given as > 5000 mg/kg body weight (no further details; Moreno, 1976).

A 20% solution of acetic acid methyl ester caused local anaesthesia within an average period of 11 minutes when applied to the outer ear of the guinea pig (Ikebe, 1934).

After inhalation of 16,000 ppm acetic acid methyl ester (nominal concentration) for 4 h, none of the exposed albino rats died, while after inhalation of 32,000 ppm (nominal concentration) for 4 h, all 6 exposed animals died. The observation period was 14 days (no further details; Smyth et al., 1962).

The threshold concentrations at which narcosis-like symptoms occurred after exposure to acetic acid methyl ester for 5 to 6 h were 34 mg/l in white mice and 56 mg/l in cats. Clinical signs of toxicity occurring after exposure included convulsions, vomiting (cats) and impairment of breathing. The body temperature in the surviving animals was reduced (by up to 2 °C). The corneal reflex could be produced within $^1/_4$ to $^1/_2$ hour of the end of exposure. Complete recovery often took several days, particularly after inhalation of higher concentrations. The findings at autopsy were not particularly characteristic in the mice. In the cats, slight pulmonary emphysema was found in several animals as well as slight reddening of the mucous membranes of the respiratory tract. As shown in Table 5, the lethal dose for acetic acid methyl ester in mice was 80 to 105 mg/l after an exposure period of 31 to 46 min. On exposure to lower concentrations, such as 34 mg/l for 216 to 312 min, delayed deaths occurred. Higher concentrations, such as 147 mg/l, were survived following shorter exposure periods (25 to 27 min). The lowest lethal concentration for cats following a 56- to 185-minute exposure was given as 67 mg/l (see Table 6; Flury and Wirth, 1933; von Oettingen, 1960).

Table 5. Acute toxicity of acetic acid methyl ester vapour (at 22 °C) in mice (2 animals/group; sex unspecified; Flury and Wirth, 1933; von Oettingen, 1960)

Concentration (mg/l)	(ppm)	Duration of exposure (min)	Signs of toxicity
24	7900	459 and 480	No signs of toxicity in one mouse, deep narcosis in second mouse after 459 min, dyspnoea, irritation; recovery after 6 h
34	11,200	216 and 312	Eye irritation, dyspnoea, narcosis after 216 and 312 min, respectively; both mice died 10 h after exposure
42	13,860	60 and 86	Eye irritation, dyspnoea, narcosis after 60 and 86 min, respectively; both mice recovered

Table 5. (continued)

Concentration (mg/l)	(ppm)	Duration of exposure min	Signs of toxicity
63	20,790	52 and 61	Eye irritation, dyspnoea, narcosis after 52 and 61 min, respectively; one mouse died immediately after exposure, the second 3 h after exposure
80	26,400	31 and 42	Eye irritation, dyspnoea, narcosis after 31 and 42 min, respectively; one mouse recovered, the second died 3 h after exposure
105	34,650	31 and 46	Eye irritation, dyspnoea, narcosis after 31 and 46 min, respectively; one mouse died immediately after exposure, the second 1.5 h after exposure
126	41,580	23 and 35	Irritant effect, dyspnoea, convulsions, narcosis after 23 and 35 min, respectively; one mouse died immediately after exposure, the second 3 min after exposure
147	48,510	25 and 27	Irritation, paralysis, convulsions, dyspnoea, pronounced slowing of breathing, narcosis after 25 and 27 min, respectively; both mice recovered
168	55,440	10 and 20	Immediate irritant effect, dyspnoea, narcosis after 10 and 20 min, respectively; both mice died in exposure chamber, pulmonary emphysema

Table 6. Acute toxicity of acetic acid methyl ester in cats (2 animals/group) exposed to a flowing gas mixture (19 to 24 °C; Flury and Wirth, 1933; von Oettingen, 1960)

Concentration (mg/l)	(ppm)	Duration of exposure (min)	Signs of toxicity
30	9900	590	Eye irritation, dyspnoea, salivation, somnolence, no narcosis; both cats recovered
56	18,480	238 and 295	Eye irritation, dyspnoea, salivation, vomiting and convulsions in one animal, narcosis after 238 and 295 min, respectively; both cats recovered
67	22,100	56 and 185	Eye irritation, dyspnoea, salivation, convulsions, narcosis after 56 and 185 min, respectively; both cats died shortly after the end of exposure; post-mortem: hyperaemia and marginal emphysema of the lungs
95	31,350	29 and 41	Eye irritation, salivation, dyspnoea, narcosis after 29 and 41 min, respectively; one cat developed convulsions after the end of exposure, death after 50 min; second cat died 4 min after exposure; post-mortem: marginal emphysema of the lungs, congestion of the trachea
106	34,980	29 and 30	Eye irritation, salivation, vomiting, dyspnoea, narcosis after 29 and 30 min, respectively, death 2 and 5 min after exposure; post-mortem: marginal emphysema of the lungs, slight pulmonary oedema
134	44,220	10 and 20	Eye irritation, salivation, dyspnoea, convulsions, narcosis after 19 and 20 min, respectively; one cat recovered, the second died 5 min after exposure; post-mortem: marginal emphysema of the lungs, heart flaccid
163	53,790	14 and 28	Eye irritation, salivation, dyspnoea, convulsions, narcosis after 14 and 28 min, respectively; both cats died 1 and 9 min after exposure; post-mortem: marginal emphysema of the lungs, heart flaccid

The lethal concentration in mice following a 10-hour exposure was given as 11,000 ppm (Browning, 1965).

The inhalation of acetic acid methyl ester at concentrations of > 10 mg/l led to a pronounced decrease in respiratory volume and frequency in almost all of the exposed rabbits, due to the irritant effect on the mucous membranes of the nose (no further details; Flury and Wirth, 1933).

A female guinea pig (weight 545 g) was exposed to a saturated vapour of acetic acid methyl ester in a bell jar. Immediately after the animal was placed in the chamber, it closed its eyes, and lacrimation was observed after 5 min. In addition, trembling of the forepaws, spasmodic breathing and disturbances of heart rhythm occurred (no further details; Duquenois and Revel, 1934).

The minimum lethal doses of acetic acid methyl ester following subcutaneous injection were 3000 mg/kg body weight in cats and between 3000 and 5000 mg/kg body weight in guinea pigs. Signs of toxicity observed were disturbances of balance, lying on the side, convulsions and accelerated breathing (Flury and Wirth, 1933).

In rats, the minimum lethal dose of acetic acid methyl ester by subcutaneous injection was 8000 mg/kg body weight (Aiazzi-Mancini et al., 1943).

Cats survived the rectal administration of acetic acid methyl ester at doses of 2500 and 5000 mg/kg body weight. Signs of toxicity included excitation, mild salivation, evacuation of the bowels with copious mucus and accelerated breathing (Flury and Wirth, 1933).

An LD_{50} value of 214 mg/kg body weight was given for acetic acid methyl ester in mammals. Data on the species and route of application were not provided (RTECS, 1993).

A group of 4 cats, 1 male (weight 3.4 kg) and 3 females (weight 2.2 to 2.5 kg), inhaled acetic acid methyl ester for 6 h/day at concentrations of 19 to 21 mg/l (6270 to 6930 ppm) for 8 days. One of the females died on day 5 of the study. Clinical effects during exposure included eye irritation, closed eyes and salivation. In the remaining 3 cats, clear changes in haematological parameters were found immediately after the last exposure, such as increases in the numbers of erythrocytes and leukocytes and in haemoglobin, and there was also an increase in neutrophilic granulocytes in one cat with a decrease in eosinophils. In a further experiment, 2 male cats (weight 2.7 to 3.0

kg) were exposed to 4.1 mg acetic acid methyl ester/l for 7 h on each of 3 days. Irritation of the eyes and salivation were observed during exposure. The cats recovered within 8 days of the end of exposure (Flury and Wirth, 1933).

7.3 Skin and mucous membrane effects

Neat acetic acid methyl ester (99.9% pure) was applied to the dorsal skin of 3 New Zealand white rabbits for 4 h under semi-occlusive cover (OECD guideline no. 404). The rabbits showed very mild erythema 30 min to 24 h after removal of the patch. 1 rabbit had dry, rough skin 24 to 72 h after removal. All of the rabbits were free of irritant effects 48 h after removal of the patch. Based on these results, acetic acid methyl ester was evaluated as not irritating (Hoechst, 1988 b).

The application of 2 ml acetic acid methyl ester to the clipped skin of rabbits caused erythema (no further details; Flury and Neumann, 1927).

Acetic acid methyl ester caused mild irritation in rabbits following 24-hour occlusive application to the intact and scarified skin (no further details; Moreno, 1976).

In a further skin irritation study (Draize test), the application of 20 mg acetic acid methyl ester for 24 h caused moderate irritation (no further details; Marhold, 1986).

Acetic acid methyl ester (0.01 ml, neat) was not irritating after 24 h when applied uncovered to the shaved abdominal skin of rabbits (grade 1 on a 10-point scale; Smyth et al., 1962).

Rabbits (3, New Zealand white) had 0.1 ml 99.9% pure acetic acid methyl ester instilled into the conjunctival sac of the left eye (OECD guideline no. 405). The eyes were evaluated 1, 24, 48 and 72 h after instillation. Swelling of the conjunctiva with white, mucus discharge, half closed eyelids with marked hyperaemia, reddened irises and diffuse clouding of the cornea were reported 1 to 72 h after instillation. 24 to 72 h after instillation, the conjunctiva was stained white and bleeding, and sections were detached. All of the rabbits were free from irritant effects 7 days after instillation. Based on these results, the substance was evaluated as irritating to the eye (Hoechst, 1988 c).

Vapour concentrations of 32 mg/l caused eye irritation and salivation in mice and rabbits (Flury and Neumann, 1927).

In a standard Draize test, 100 mg acetic acid methyl ester caused moderate irritation to the rabbit eye after 24 h (no further details; Marhold, 1986).

Following instillation of 0.1 ml 98% pure acetic acid methyl ester into the eyes of 4 rabbits, a maximum average score of 39.50 was reported. The substance was thus moderately irritating (no further details; Bagley et al., 1992).

In the rabbit eye, 0.005 ml neat acetic acid methyl ester caused grade 5 irritation on a 10-point scale (Smyth et al., 1962).

The instillation of one drop of acetic acid methyl ester into the conjunctival sac of rabbits caused chemosis similar to that caused by 2% acetic acid. In all cases, the irritation cleared up within 1 to 2 days, without leaving lasting damage to the eye (Flury and Wirth, 1933).

In a Draize test involving the instillation of 0.1 ml neat acetic acid methyl ester (purity unspecified) into the eyes of 4 rabbits, the eyes were assessed 24, 48 and 72 h, 7, 10, 14 and 21 days after instillation. After 24 h, a score of 39 out of 110 was determined and the substance was evaluated as moderately irritating. The thickness of the cornea was also measured in the same rabbits at each time point, with the aid of a split lamp microscope (3 measurements/cornea). The average corneal thickness before instillation was 0.25 ± 0.03 mm, while 24 h after instillation, the cornea was 70% thicker than normal. According to the authors, there was a significant linear correlation between the Draize score and the thickness of the cornea, and thus changes in corneal thickness could be used to quantify the irritant effect (Kennah et al., 1989 a).

A moderate irritant effect was also found in a further in vitro eye irritation test. In this test, bovine corneas were fixed in special 2-chamber holders and incubated at 32 °C for one hour. The epithelial side of the cornea was then incubated with 0.5 ml acetic acid methyl ester at concentrations of 20, 40, 60, 80 or 100% for 30 min at room temperature (6 corneas/concentration). The epithelial side was then washed and incubated for a further 2 h. The opacity of the cornea was measured after this recovery period, and was 27.8 at a concentration of 20%, 43.7 at 40%, 40.8 at 60%, 42.0 at 80% and 44.8 at 100% (controls 2.8). The authors described 0 to 20 opacity units as slightly irritating, 21 to 40 as slightly to moderately irritating, 41 to 70 as moderately irritating and 71 and over as severely irritating (Gautheron et al., 1992).

7.4 Sensitization

No information available.

7.5 Subchronic and chronic toxicity

No information available.

7.6 Genotoxicity

7.6.1 In vitro

The mutagenic potential of acetic acid methyl ester was tested in an Ames test in strains TA 100, TA 1535, TA 1537, TA 1538 and TA 98 and in *Escherichia coli* strain WP2uvrA. The test was carried out with the plate incorporation method and the substance was 99.9% pure. An Aroclor 1254-induced rat liver S9 enzyme preparation was used as an in vitro metabolizing agent. The concentrations of acetic acid methyl ester tested ranged from 4 to 5000 μg/plate, and the solvent used was DMSO. Acetic acid methyl ester did not induce reversions in any of the strains of bacteria tested and was thus not mutagenic in this Ames test (Hoechst, 1988 d).

Acetic acid methyl ester was tested in an Ames test in strains TA 100, TA 1535, TA 1537, TA 97 and TA 98 under the US National Toxicology Program (NTP). The preincubation method was used with and without S9-mix (Aroclor 1254-induced rat and Syrian hamster liver enzymes). The substance tested was 99% pure, and the study was carried out in accordance with an established protocol. The concentrations of acetic acid methyl ester tested ranged from 100 to 10,000 μg/plate, the medium used for preparing the dilutions was water. Acetic acid methyl ester did not induce reversion in any of the Salmonella typhimurium strains tested. Even after treatment at the highest concentration, all of the colony frequencies were within the range of the respective strain-specific spontaneous reversion frequency. Thus in this test, no mutagenic potential of acetic acid methyl ester could be established (Zeiger et al., 1992).

Along with a series of other substances (various ketones and organic acid esters), acetic acid methyl ester was tested for its potential to induce mitotic chromosomal malsegregation, mitotic recombination and point mutations in *Saccharomyces cerevisiae* strain D61.M. The purity of the substances tested, including acetic acid methyl ester, was > 97%. Appropriate amounts of the liquid substances were pipetted directly into the medium. The yeast cells were treated with the substances for 4 h at 28 °C, then for 16 h or more at 0 °C (on ice) followed by a further 4 h at 28 °C. Acetic acid methyl ester clearly induced aneuploidy in this test system, but no mitotic recombination or gene mutations. With the treatment scheme described, more aneuploid colonies formed in acetic acid methyl ester treated cells than in negative control cells by a factor of 185. Therefore acetic acid methyl ester should be evaluated as a substance with strong potential to induce aneuploidy in the *Saccharomyces cerevisiae* D61.M test system (Zimmermann et al., 1985).

Another study investigated the metabolic activation of cyclophosphamide and the mutagenic activity connected with this in *Saccharomyces cerevisiae* (strains D_4 and SP198) under various culture conditions. The metabolic activation was induced with S9 (phenobarbital- and β-naphthoflavone-treated liver from rats) and cofactors. The in vitro effect of acetic acid methyl ester on the genotoxic activity of cyclophosphamide was tested in strain D_4, with the criterion tested being gene conversion. A non-toxic concentration of acetic acid methyl ester was used in the incubation medium, 5% v/v. The cells were suspended in phosphate buffer. Acetic acid methyl ester (> 99.5% pure) caused a significant reduction in the induction of gene conversion by cyclophosphamide (5 mM) at the adenine (ade 2) and tryptophan (trp) loci. The conclusion from these results is that acetic acid methyl ester can quantitatively change the induction of mutational events, if it is used as a solvent vehicle in studies with substances which, in vitro, must be treated with metabolizing agents (e.g. S9-mix) to be activated. The possible mutagenic properties of acetic acid methyl ester alone were not mentioned (Barale et al., 1983).

7.6.2 In vivo

No information available.

7.7 Carcinogenicity

No information available.

7.8 Reproductive toxicity

No information available.

7.9 Effects on the immune system

No information available.

7.10 Neurotoxicity

Wistar rats (weight ca. 260 g; sex and number unspecified) inhaled a varnish thinner at a concentration of 10,000 ppm which, besides toluene (52.8%) and isopropanol (21.7%), also contained acetic acid methyl ester (12.6%). The rats were exposed to the thinner in a plastic container twice daily for 10 min, 6 days/week for 12 and 14 months. Data on treatment of control rats were not provided. Reduced body weight gain was observed. At the end of the study, the cerebral cortices from 4 control rats, 3 rats that had been exposed for 12 months and 12 rats that had been exposed for 14 months were examined by light- and electron microscopy. Compared with the controls, the exposed rats showed an increase in cristae and deformed cristae in the mitochondria in neurones and axons, as well as an increase in endoplasmic reticulum and ribosomes and an enlarged Golgi apparatus in the neurones. These effects were interpreted by the authors as active cellular adaptation. A significant increase in lysosomes and lipid material was also found in the neurones, an effect which was judged to be a degenerative process (no further details; Hashimoto et al., 1983).

7.11 Other effects

In an in vitro cytotoxicity test in BALB/c 3T3 murine fibroblasts to predict the eye irritation potential of acetic acid methyl ester, only a low correlation between the concentration of the substance which inhibited the normal growth rate by 50% (GI_{50}; growth inhibition) and the thickness of the cornea after exposure (see Sect. 7.3; Kennah et al., 1989 a) was found (Kennah et al., 1989 b).

The effect of acetic acid methyl ester on cell survival was studied at concentrations of from 0.5 to 10% in Schizosaccharomyces pombe (P_1-strain) and in Chinese hamster V79 cells. Acetic acid methyl ester was of comparatively low toxicity. The survival rate of Schizosaccharomyces pombe on exposure for 24 h was unimpaired at all tested concentrations. In contrast, the survival rate in V79 cells cultured as isolated single cells (low cell density) was reduced at all concentrations tested, while that in cells cultured as monolayers (high cell density) was decreased at concentrations of 5% and above (Abbondandolo et al., 1980).

The activity of aminopyrine demethylase from phenobarbital-induced and non-induced mouse liver microsomes was markedly reduced by acetic acid methyl ester at a concentration of 5% (Abbondandolo et al., 1980).

Acetic acid methyl ester was included in a study to test substances for their ability to increase skin permeability for the transdermal application of pharmaceuticals. Excised rat skin (6 cm^2) from the abdominal region was placed in a Franz diffusion cell. The donor phase (ca. 5 ml) consisted of indomethacin dissolved in acetic acid methyl ester, the receptor solution was isotonic saline. The incubation period was 72 h. The indomethacin concentration in the receptor phase was determined by means of HPLC every 2 h. When acetic acid methyl ester was used as the solvent, 240 mg indomethacin/cm^2/hour penetrated the skin, while when water was used, only 0.15 mg indomethacin/cm^2/hour penetrated the skin (1600-fold increase due to acetic acid methyl ester; Catz and Friend, 1989).

Following incubation of 50 µl of a 0.5 M solution of acetic acid methyl ester in ethanol with S9-homogenate from the nasal tissues of rats, the hydrolysis rate was 15 ± 3 nmol/mg S9-protein/minute (Dahl et al., 1987).

In contrast to 2-butanone and 2,5-hexanedione, acetic acid methyl ester (0.625 to 2.5% v/v) did not stimulate the activity of (Ca^{2+}/Mg^{2+}) ATPase in synaptosomal membranes from rats in vitro (Edelfors and Ravn-Jonsen, 1992).

Acetic acid methyl ester inhibited the assembly of porcine brain tubulin in vitro (Gröschel-Stewart et al., 1985).

The concentration of acetic acid methyl ester that inhibited the haemolysis of rat erythrocytes in hypotonic medium by 100% was 5000 ppm (Holmberg et al., 1974).

A concentration of 8500 mg acetic acid methyl ester/l (115 mM) was given as that which haemolysed bovine erythrocytes in vitro (Fühner and Neubauer, 1907).

Low concentrations of acetic acid methyl ester have a negative inotropic effect on the isolated mammalian heart (Aiazzi-Mancini et al., 1943).

8. Experience in humans

No reports of industrial experience or cases of poisoning with pure acetic acid methyl ester are available. The ester commonly occurs as a component of solvent mixtures. The majority of publications concern case reports, in which data on concentration and duration of exposure can often not be determined retrospectively. Frequently, intoxications due to mixtures are involved, and these do not permit a clear evaluation of the toxicity of acetic acid methyl ester.

Exposure to acetic acid methyl ester by inhalation causes drowsiness and long-lasting headaches in man (no further details; Flury and Wirth, 1933).

Female workers in a shoe factory, who were exposed to the vapours of a mixture of formic acid methyl ester, formic acid ethyl ester, acetic acid methyl ester and acetic acid ethyl ester (no further details) reported eye irritation, visual disturbances, heart and respiratory complaints and nervous disturbances, such as intoxicated states and depressive moods (Duquenois and Revel, 1934).

The inhalation of paint thinner (components: acetic acid esters including the methyl ester) caused intoxication and hallucinations in man (Tambo, 1973).

An early report recorded a case of bilateral atrophy of the optic nerve in a 69-year-old man following inhalation exposure to acetic acid methyl ester in the workplace. The man first felt unwell with giddiness, headaches, exhaustion, vertigo and an unsteady gait. During one of these attacks, he suddenly became blind in both eyes. In the course of a few days, his sight returned, but again deteriorated severely one month later. During a stay in hospital, his sight improved somewhat. A large central scotoma was noted in the left eye, and colours could no longer be discerned with the right eye. Bilateral atrophy of the optic nerve was established 4 months after admission to hospital. The author attributed the optic nerve damage to the formation of formaldehyde or to a direct effect of the methanol formed (Lund, 1944).

Blindness due to atrophy of the optic nerve (optic neuropathy) occurred in a 17-year-old girl who had sniffed varnish thinner over a period of 3 months. The main components detected in the vapour phase of the thinner by gas chromatography were methanol and acetic acid methyl ester (Kohriyama et al., 1989).

A 21-year-old man, who had habitually sniffed glue and paint thinner from the age of 15, suffered increasing loss of central visual acuity progressing to complete blindness, as well as dysesthesia of the feet and weakness of the legs, from the fourth day after he had started sniffing a new paint thinner. His visual acuity improved within ca. 3 months of the first occurrence of visual disturbance, colour discrimination was completely restored and the dysesthesia of the feet had completely gone. In contrast to the normal paint thinner (15 vol.-% methanol, 11 vol.-% ethyl acetate, 67 vol.-% toluene, 5 vol.-% isobutanol, 2 vol.-% cellosolve), the new paint thinner, in addition to methanol (20 vol.-%) consisted mainly of acetic acid methyl ester (20 vol.-%) and toluene (57 vol.-%). As sniffing the previously used paint thinner for several years had not led to visual disturbances, the occurrence of visual disturbances at this time was attributed to the combination of acetic acid methyl ester and methanol (20 vol.-% each) in the new paint thinner (Ogawa et al., 1988).

Workers who were exposed to acetic acid methyl ester vapour and who at times wore breathing masks, had comparable amounts of methanol in their urine at the end of the working day (4.8 mg/l) to workers who were exposed to methanol (no further details; Elkins, 1954).

To test the process of assessing acetic acid methyl ester and methanol exposure based on the determination of the methanol concentration in the urine of exposed subjects, 2 male volunteers (aged 33 and 48 years) were repeatedly exposed to levels close to the maximum workplace concentration of this solvent (200 ppm), at a temperature of 20 to 25 °C, in a chamber of ca. 18 m^3, in which the air could circulate. The relationship between the air concentrations of acetic acid methyl ester and methanol and methanol excretion in the urine was examined. The concentrations of the substances in the air were measured by colorimetry and gas chromatography, while the methanol concentration in the urine was determined by the hot separation chromotropic acid method. The base levels of methanol in the urine were determined before the study. In healthy subjects not exposed to methanol, the methanol concentrations in the urine (specific gravity 1.010 to 1.053) ranged from 1 to 5 mg/l. The intake of vitamins or ethanol had no effect on the methanol levels in the urine. The methanol concentrations in the urine of 12 workers from a papering and painting business, who did not come into contact with methanol or acetic acid methyl ester, were determined before and after work. The methanol concentrations in 24-hour urine samples ranged from 1 to 5 mg/l. The average value and standard deviation was 2.95 ± 0.01 mg/l. The urine of the two male volunteers was tested for its methanol content every 8 hours for 12 days before the beginning of the study. The methanol concentrations standardized against a urine with a specific gravity of 1.024 were between 1 and 5 mg/l. The individual average value and standard deviation for one subject, who drank a lot of water, was 2.03 ± 0.92, and that for the second subject was 2.92 ± 0.99 mg/l. Acetic acid methyl ester or methanol vapours were mixed with the circulating air in the exposure chamber at concentrations of about 200 ppm. The volunteers worked while sitting down in the chamber, and every 15 min carried out concentration measurements. The exposure was carried out twice daily (from 10.00 to 12.00 and from 14.00 to 16.00 h). The concentration of the solvent in the breath was measured immediately after the end of exposure at defined time intervals for up to 2 h. In addition, the methanol concentration was determined in the 2-, 4- and 8-hour urine. The two exposure studies with acetic acid methyl ester were carried out on 3 consecutive days, those with methanol on 3 and 4 consecutive days. In all studies, the amount of methanol in the urine increased during the 2-hour exposure in the morning and reached a maximum during the 2-hour exposure in the afternoon and 2 to 4 h afterwards. The amount of methanol then decreased again and mainly reached the normal level of < 5 mg/l before the next exposure on the following day. Because of large fluctuations in the acetic acid methyl ester concentrations, the methanol concentration in the urine varied more markedly in the studies with this solvent. However, the unmetabolized ester was only analyzed in the breath and not in the urine. The value was 5 to 20 ppm 15 min after the end of the acetic acid methyl ester exposure. In contrast to methanol, no clear results were obtained for the excretion of acetic acid methyl ester in the breath. It was not possible to elucidate the course of excretion. The authors indicated that a tendency was evident for the inhaled acetic acid methyl ester to be metabolized to methanol in the body and to be expired as such (see Table 7).

Table 7. Exposure studies with methanol and acetic acid methyl ester in two male volunteers (Tada et al., 1974)

	Methanol		Acetic acid methyl ester	
	Study I	Study II	Study I	Study II
Average exposure concentration in ppm				
By colorimetry	192	199	220	210
By gas chromatography	192	199	223	198
Methanol concentration in the urine in mg/l				
Peak concentration (in each case after the afternoon exposure)	9.2–13.8	7.0-23.2	8.3–21.6	8.3-22.6
Average concentration	11.5	11.9	13.6	14.8
Solvent concentration in the breath in ppm				
Immediately after exposure	30 – 35		no data	
15 min after exposure	20		5 – 20	
30 min after exposure	15		no data	
60 min after exposure	< 1		no data	

Based on these study results, a urinary methanol concentration of > 10 mg/l, measured in workers before the end or within 4 h of the end of work, is considered as a sign that the maximum permitted workplace concentration of 200 ppm has been exceeded (Tada et al., 1974).

A study was carried out in 47 Swedish paint industry workers, 38 of whom (33 men, 5 women; 27 to 64 years old) had been exposed for more than 10 years to a range of solvents (including xylene, toluene, isobutanol, n-butanol, the methyl, ethyl and butyl esters of acetic acid, methylene chloride, methylethylketone and isopropanol), and 9 of whom (all men; 20 to 58 years old) were exposed to high solvent concentrations during cleaning of the tanks (independent of duration of employment). No changes in liver enzyme activities (alanine aminotransferase, aspartate aminotransferase, γ-glutamyl transferase, ornithine carbamyltransferase) were observed in the serum compared with control subjects (47 non-exposed workers, similar to the exposed workers with respect to age, sex, place of residence and consumption of tobacco, alcohol and medicines). The solvent-exposed workers performed less well than the control group in tests of reaction time, speed of perception and short-term memory before and after work. There were no differences in the "critical flicker fusion" test. The highly exposed workers showed even worse results than the workers exposed for long periods, and thus impairment of CNS function appears to be associated with acute solvent exposure. The sister chromatid exchange rate (17 subjects tested) and the chromosome aberration rate (5 subjects tested) were not increased in the lymphocytes of the workers exposed to particularly high concentrations compared with the controls. Each worker was exposed to between 3 and 9 different solvents at the same time. Of the 47 workers, 11 were exposed to solvents including on average 13 mg acetic acid methyl ester/m^3 air (3 to 169 mg/m^3). The average acetic acid methyl ester concentration during a work shift (8 h) was ca. 6 mg/m^3/day for workers exposed long-term (n=7) and ca. 69 mg/m^3/day for highly-exposed workers (n=4; Anshelm Olson, 1982; Haglund et al., 1980; Lundberg and Hakansson, 1985). This mixed exposure does not permit a clear evaluation of the toxicity of acetic acid methyl ester.

In studies of 6 furniture polishers, who were suffering from so-called polishers' keratitis (numerous minute round, oval or irregularly shaped corneal vacuoles, conjunctiva and eyelids that were barely or slightly red, photophobia), toluene and the methyl, ethyl and butyl esters of acetic acid (components of the thinner solution in the polish) were held to be responsible for the effects on the cornea (Schmid, 1956).

In a study in a male volunteer who was exposed to 165 to 290 ppm acetic acid methyl ester for 4 h/day over 4 days in an 18 m^3 inhalation chamber, there were no significant differences from the control in tests of time perception (estimation of 5, 10 and 30 s) during exposure (each carried out every 30 min; Nakaaki, 1974).

The application of 2 drops of acetic acid methyl ester to human skin caused cooling and flaking but no irritation (no further details; Reus, 1933).

Acetic acid methyl ester (10% in vaseline) was not irritating in 48-hour patch tests in man (no further details; Kligman, 1976).

A maximization test with acetic acid methyl ester (10% in vaseline) caused no sensitization reaction in 25 volunteers (Kligman, 1976).

In 2 of 4 volunteers (no detailed information), inhalation exposure to acetic acid methyl ester (1, 15 and 30 mg/l) for 5 min in an inhalation chamber (10 m^3), caused a feeling of coldness and dryness in the throat and mild irritation of the throat and trachea at 15 mg/l and above. The irritant effects disappeared almost immediately on leaving the study room (Flury and Wirth, 1933).

The application of one drop of acetic acid methyl ester to the human tongue caused an immediate burning sensation with subsequent reddening and swelling (Reus, 1933).

The olfactory threshold for acetic acid methyl ester in the air has been given as 4.6 ppm. At a concentration of 200 ppm, 50 to 90% of volunteers were able to detect the smell of acetic acid methyl ester in the air (number of volunteers not specified; Amoore and Hautala, 1983).

In a further study, the olfactory threshold for acetic acid methyl ester in the air was given as 200 ppm (14 volunteers; at 20 °C, 1013 hPa; May, 1966).

Considerable individual variation in the ability to detect the smell of acetic acid methyl ester in the air has been found. The lower olfactory threshold value (odour perception by at least one volunteer) was 610 mg/m^3 (equivalent to 200 ppm) and the upper olfactory threshold value (odour perception by all volunteers) was 915 mg/m^3

(equivalent to 300 ppm; number of volunteers not specified). The irritant concentration was given as 30,496.9 mg/m^3 (Ruth, 1986).

A further olfactory threshold for acetic acid methyl ester in the air was given as 0.5 mg/m^3, the threshold value for changes in light sensitivity of the eyes as 0.18 mg/m^3 and that for effects on "electrical activity" in the brain as 0.08 mg/m^3 (Hofmekler, 1960).

9. Threshold limit values

The MAK value in the Federal Republic of Germany is 200 ml/m^3, or 610 mg/m^3. The substance is classified in pregnancy group D (DFG, 1994; TRGS, 1994).

In the USA, the TLV value is also 200 ml/m^3, or 606 mg/m^3 (ACGIH, 1993).

References

Abbondandolo, A., Bonatti, S., Corsi, C., Corti, G., Fiorio, R., Leporini, C., Mazzaccaro, A., Nieri, R.
The use of organic solvents in mutagenicity testing
Mutat. Res., 79, 141–150 (1980)

ACGIH (American Conference of Governmental Industrial Hygienists, Inc.)
Threshold limit values for chemical substances and physical agents and biological exposure indices 1993–1994 (1993)

Aguilo, A., Hobbs, C.C., Zey, E.G.
Acetic acid
In: Ullmann's encyclopedia of industrial chemistry
5th ed., vol. A1, p. 59
VCH Verlagsgesellschaft, Weinheim (1985)

Aiazzi-Mancini, M., Noferi, G., Schilling, G., Cambi, L.
Ricerche farmacologiche e tossicologiche sull'acetato di metile e sull'acetato di etile
Boll. Soc. Ital. Biol. Sper., 18, 45–47 (1943)

Amoore, J.E., Hautala, E.
Odor as an aid to chemical safety: odor thresholds compared with threshold limit values and volatilities for 214 industrial chemicals in air and water dilution
J. Appl. Toxicol., 3, 272–290 (1983)

Anshelm Olson, B.
Effects of organic solvents on behavioral performance of workers in the paint industry
Neurobehav. Toxicol. Teratol., 4, 703–708 (1982)

Bagley, D.M., Botham, P.A., Gardner, J.R., Holland, G., Kreiling, R., Lewis, R.W., Stringer, D.A., Walker, A.P.
Eye irritation: reference chemicals data bank
Toxicol. in Vitro, 6 (6), 487–491 (1992)

Barale, R., Rusciano, D., Stretti, G., Zucconi, D., Monaco, M., Mosesso, P., Principe, P.
The induction of forward gene mutation and gene conversion in yeasts by treatment with cyclophosphamide in vitro and in vivo
Mutat. Res., 111, 295–312 (1983)

Browning, E.
Toxicity and metabolism of industrial solvents, p. 526–528
Elsevier Publishing Co., Amsterdam (1965)

Budavari, S., O'Neil, M.J., Smith, A., Heckelman, P.E. (eds.)
The Merck index
11th ed., p. 948
Merck & Co., Rahway, New Jersey (1989)

Catz, P., Friend, D.R.
Alkyl esters as skin permeation enhancers for indomethacin
Int. J. Pharm., 55, 17–23 (1989)

Dahl, A.R., Miller, S.C., Petridou-Fischer, J.
Carboxylesterases in the respiratory tract of rabbits, rats and Syrian hamsters
Toxicol. Lett., 36, 129–136 (1987)

DFG (Deutsche Forschungsgemeinschaft, Kommission zur Prüfung gesundheitsschädlicher Arbeitsstoffe)
MAK- und BAT-Werte-Liste 1994, Mitteilung 30
VCH Verlagsgesellschaft, Weinheim (1994)

Duquenois, P., Revel, P.
Intoxications professionnelles par vapeurs de quelques esters employés comme solvants
J. Pharm. Chim., 19, 590–601 (1934)

Edelfors, S., Ravn-Jonsen, A.
Effect of organic solvents in nervous cell membrane as measured by changes in the (Ca^{2+}/Mg^{2+})ATPase activity on fluidity of synaptosomal membrane
Pharmacol. Toxicol., 70, 181–187 (1992)

Elkins, H.B.
Analyses of biological materials as indices of exposure to organic solvents
Arch. Ind. Hyg. Occup. Med., 9, 212–222 (1954)

Falbe, J., Regitz, M. (eds.)
Römpp Chemie Lexikon
9th ed., vol. 2, p. 1235
Georg Thieme Verlag, Stuttgart, New York (1990)

Filov, V.A.
Retention und Umwandlung von nieder molekularen Estern im Organismus (German translation of the Russian)
Gig. Tr. Prof. Zabol., 5, 14 (1961)

Filov, V.A.
Kinetics of accumulation of substances in biological systems characterized by free penetration and subsequent breakdown (English translation of the Russian)
Dokl. Akad. Nauk. SSSR., 157, 612–615 (1964 a)

Filov, V.A.
Über den Vergleich der Intensität der narkotischen Wirkung von Methyl- und Ethylacetat sowie ihrer Metabolismusprodukte (German translation of the Russian)
Farmakol. Toksikol., 27, 492–493 (1964 b)

Flury, F., Neumann, W. (1927)
Cited in: von Oettingen (1960)

Flury, F., Wirth, W.
Zur Toxikologie der Lösungsmittel. (Verschiedene Ester, Aceton, Methylalkohol)
Arch. Gewerbepathol. Gewerbehyg., 5, 1–90 (1933)

Führer, H., Neubauer, E.
Hämolyse durch Substanzen homologer Reihen
Arch. Exp. Pathol. Pharmakol., 56, 333–345 (1907)

Gautheron, P., Dukic, M., Alix, D., Sina, J.F.
Bovine corneal opacity and permeability test: an in vitro assay of ocular irritancy
Fundam. Appl. Toxicol., 18, 442–449 (1992)

Ghittori, S., Imbriani, M., Borlini, F., Pezzagno, G., Zadra, P.
Studio sulla stabilita degli esteri nel sangue in vitro
Boll. Soc. Ital. Biol. Sper., 60 (11), 2207–2213 (1984)

Gröschel-Stewart, U., Mayer, V.W., Taylor-Mayer, R.E., Zimmermann, F.K.
Aprotic polar solvents inducing chromosomal malsegregation in yeast interfere with the assembly of porcine brain tubulin in vitro
Mutat. Res., 149, 333–338 (1985)

Haglund, U., Lundberg, I., Zech, L.
Chromosome aberrations and sister chormatid exchanges in Swedish paint industry workers
Scand. J. Work Environ. Health, 6, 291–298 (1980)

Hashimoto, S., Maeda, M., Hiruma, S., Satou, T., Sakatani, K.
A study of chronic thinner intoxication with special reference to pathomorphological changes of the cerebral cortex of experimental animals (Japanese with English abstract)
Kankyo Kagaki, Kenkyusho Kenkyu Hokoku (Kinki Daigaku), 11, 217–223 (1983)

Hoechst AG, Pharma Forschung Toxikologie und Pathologie
Methylacetat – Prüfung der akuten dermalen Toxizität an der Wistar-Ratte
Unpublished report no. 88.0181 (1988 a)

Hoechst AG, Pharma Forschung Toxikologie und Pathologie
Methylacetat – Prüfung auf Hautreizung am Kaninchen
Unpublished report no. 88.0125 (1988 b)

Hoechst AG, Pharma Forschung Toxikologie und Pathologie
Methylacetat – Prüfung auf Augenreizung am Kaninchen
Unpublished report no. 88.0168 (1988 c)

Hoechst AG, Pharma Research Toxicology and Pathology
Methylacetat – Study of the mutagenic potential in strains of *Salmonella typhimurium* (Ames test) and *Escherichia coli*
Unpublished report no. 88.0406 (1988 d)

Hoechst AG
EUCLID data set Methyl acetate (1994 a)

Hoechst AG
EC safety data sheet Methylacetat (1994 b)

Hofmekler, V.A.
Data to substantiate the maximum permissible concentration of acetates in the atmosphere (Russian with English abstract)
Gig. Sanit., 25, 9–15 (1960)

Holmberg, B., Jakobson, I., Malmfors, T.
The effect of organic solvents on erythrocytes during hypotonic hemolysis
Environ. Res., 7, 193–205 (1974)

Ikebe, S.
Fol. Pharmacol. Japon., 19, 62 (1934)
Cited in: von Oettingen (1960)

Kennah II, H.E., Hignet, S., Laux, P.E., Dorko, J.D., Barrow, C.S.
An objective procedure for quantitating eye irritation based upon changes of corneal thickness
Fundam. Appl. Toxicol., 12, 258–268 (1989a)

Kennah II, H.E., Albulescu, D., Hignet, S., Barrow, C.S.
A critical evaluation of predicting ocular irritancy potential from an in vitro cytotoxicity assay
Fundam. Appl. Toxicol., 12, 281–290 (1989b)

Kligman, A.M.
Report to RIFM no. 76–191, 18 May (1976)
Cited in: Opdyke (1979)

Kohriyama, K., Hori, H., Murai, Y., Ninomiya, H., Tsukamoto, Y.
Optic neuropathy induced by thinner sniffing (Japanese with English abstract)
J. UOEH, 11, 449 - 453 (1989)

Leahy, D.E.
Intrinsic molecular volume as a measure of the cavity term in linear solvation energy relationships: octanol-water partition coefficients and aqueous solubilities
J. Pharm. Sci., 75, 629 – 636 (1986)

Leo, A., Hansch, C., Elkins, D.
Partition coefficients and their uses
Chem. Rev., 71, p. 525–555, 560, 613–616 (1971)

Lund, A.
Toxic amblyopia after inhalation of methyl acetate
Ugeskr. Laeg., 106, 308–311 (1944)

Lundberg, I., Hakansson, M.
Normal serum activities of liver enzymes in Swedish paint industry workers with heavy exposure to organic solvents
Br. J. Ind. Med., 42, 596–600 (1985)

Marhold, J.
Prehled prumyslove toxikologie; organicke latky, 353 (1986)
Cited in: RTECS (1993)

May, J.
Geruchsschwellen von Lösemitteln zur Bewertung von Lösemittel-gerüchen in der Luft
Staub-Reinh. Luft, 26, 385–389 (1966)

Moreno, O.M.
Report to RIFM, 12 May (1976)
Cited in: Opdyke (1979)

Munch, J.C.
Aliphatic alcohols and alkyl esters: narcotic and lethal potencies to tadpoles and to rabbits
Ind. Med., 41 (4), 31–33 (1972)

Nakaaki, K.
An experimental study on the effect of exposure to organic solvent vapor in human subjects
J. Sci. Lab., 50 (2), 89–96 (1974)

Oettingen, von, W.F.
The aliphatic acids and their esters: toxicity and potential dangers. The saturated monobasic aliphatic acids and their esters
Arch. Ind. Health, 21, 40–77 (1960)

Ogawa, Y., Takatsuki, R., Uema, T., Seki, Y., Hiramatsu, K., Okayama, A., Morimoto, K., Kaneko, Z.
Acute optic neuropathy induced by thinner sniffing: inhalation of mixed organic solvent containing methyl alcohol and methyl acetate
Ind. Health, 26, 239–244 (1988)

Opdyke, D.L.J.
Monographs on fragrance raw materials: methyl acetate
Food Cosmet. Toxicol., 17, 859–861 (1979)

Reus, K.J.
Inaugural Dissertation, Würzburg (1933)
Cited in: von Oettingen (1960)

RTECS (Registry of Toxic Effects of Chemical Substances)
Acetic acid, methylester, AI9100000
Produced by NIOSH (National Institute for Occupational Safety and Health) (1993)

Ruth, J.H.
Odor thresholds and irritation levels of several chemical substances: a review
Am. Ind. Hyg. Assoc. J., 47, A142–A151 (1986)

Schmid, E.
Die Augenhornhauterkrankung der Möbelpolierer
Arch. Gewerbepathol. Gewerbehyg., 15, 37–44 (1956)

Smyth, H.F., jr., Carpenter, C.P., Weil, C.S., Pozzani, U.C., Striegel, J.A.
Range-finding toxicity data: list VI
Am. Ind. Hyg. Assoc. J., 23, 95–107 (1962)

Tada, O., Nakaaki, K., Fukabori, S.
Verfahren zur Beurteilung der Methanol- und Methylacetat-Exposition (German translation of the Japanese)
J. Sci. Lab., 50, 239–248 (1974)

Tambo, S.
Toxicity hazards of paint thinners with particular emphasis on the metabolism and toxicity of acetate esters (English translation of the Japanese)
Nichidai Igaku Zasshi, 32, 349–360 (1973)

TRGS (Technische Regeln für Gefahrstoffe) 900
Grenzwerte in der Luft am Arbeitsplatz
Bekanntmachung des Bundesministers für Arbeit und Sozialordnung
Bundesarbeitsblatt 6/1994, p. 34–53

Verschueren, K. (ed.)
Handbook of environmental data on organic chemicals
2nd ed., p. 827–828, 1229, 1234–1235, 1239
Van Nostrand Reinhold, New York (1983)

Weast, R.C. (ed.)
CRC Handbook of chemistry and physics
69th ed., p. C-48
CRC Press Inc., Boca Raton, Florida (1988)

Zeiger, E., Anderson, B., Haworth, S., Lawlor, T., Mortelmans, K.
Salmonella mutagenicity tests: V. Results from the testing of 311 chemicals
Environ. Mol. Mutagen., 19, Suppl. 21, 2–141 (1992)

Zimmermann, F.K., Mayer, V.W., Scheel, I., Resnick, M.A.
Acetone, methyl ethyl ketone, ethyl acetate, acetonitrile and other polar aprotic solvents are strong inducers of aneuploidy in *Saccharomyces cerevisiae*
Mutat. Res., 149, 339–351 (1985)

7-Amino-4-hydroxy-2-naphthalenesulfonic acid

1. Summary and assessment

7-Amino-4-hydroxy-2-naphthalenesulfonic acid is of very low toxicity on acute oral administration (LD_{50} rat oral >5000 mg/kg body weight).

In rabbits, 7-amino-4-hydroxy-2-naphthalenesulfonic acid is not irritating to the intact skin or the eye.

7-Amino-4-hydroxy-2-naphthalenesulfonic acid of high purity is not mutagenic in the Salmonella/microsome test.

2. Name of substance

2.1	Usual name	7-Amino-4-hydroxy-2-naphthalenesulfonic acid
2.2	IUPAC-name	7-Amino-4-hydroxynaphthalene-2-sulfonic acid
2.3	CAS-No.	87–02–5
2.4	EINECS-No.	201–718–9

3. Synonyms, common and trade names

2-Amino-5-hydroxynaphthalene-7-sulfonic acid
2-Amino-5-naphthol-7-sulfonic acid
6-Amino-1-hydroxy-3-sulfonaphthalene
6-Amino-1-naphthol-3-sulfonic acid
Aminonaphthol sulfonic acid J

I acid
Isogamma acid
J acid
7-Amino-4-hydroxy-2-naph-thalinsulfonsäure

4. Structural and molecular formulae

4.1 Structural formula

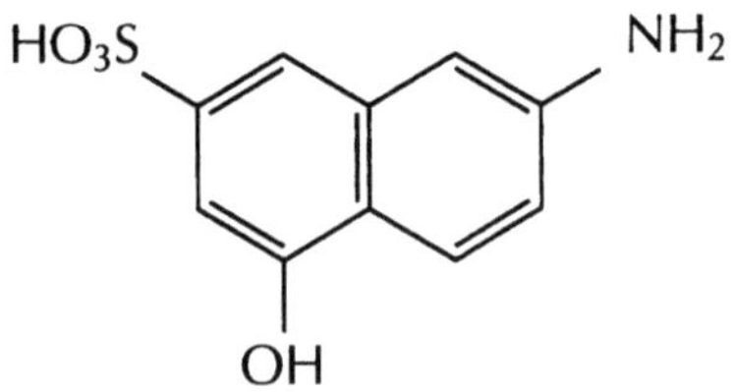

4.2 Molecular formula $C_{10}H_9NO_4S$

5. Physical and chemical properties

5.1	Molecular mass, g/mol	239.25
5.2	Melting point, °C	180 – 200 (decomposition) (Bayer, 1990)
5.3	Boiling point, °C	No information available
5.4	Vapour pressure, hPa	No information available
5.5	Density, g/cm3	ca. 0.7 (bulk density) (Bayer, 1990)
5.6	Solubility in water	ca. 0.5 g/l (at 20 °C) (Bayer, 1990) ca. 1.5 g/l (at 60 °C) (Bayer, 1989)
5.7	Solubility in organic solvents	No information available
5.8	Solubility in fat	No information available

5.9 pH value — ca. 3 (at 0.5 g/l water) (Bayer, 1989)

5.10 Conversion factor — 1 ml/m³ (ppm) ≙ 9.77 mg/m³
1 mg/m³ ≙ 0.10 ml/m³ (ppm)
(at 25 °C and 1013 hPa)

6. Uses

Coupling component for numerous azo dyes (Behre, 1979).

7. Experimental results

7.1 Toxicokinetics and metabolism

No information available.

7.2 Acute and subacute toxicity

The following LD_{50} values have been determined after acute oral exposure:

Rat, male	> 5000	mg/kg body weight (Bayer, 1978)
Rat	11,500	mg/kg body weight (Marhold, 1986)

No signs of toxicity occurred in the rat on oral administration of 5000 mg/kg body weight. A group of 10 animals was used and the observation period was 14 days (Bayer, 1978).

7.3 Skin and mucous membrane effects

An aqueous preparation (ca. 500 mg 7-amino-4-hydroxy-2-naphthalenesulfonic acid/animal) was applied to the inner surface of the ears of 2 rabbits (3 to 4 kg) under an adhesive dressing for 24 h and was washed off at the end of the exposure period with water and soap/vegetable oil. The observation period was 7 days. No skin irritation was observed (Bayer, 1979).

In an eye irritation study in rabbits, the instillation of 500 mg 7-amino-4-hydroxy-2-naphthalenesulfonic acid into the conjunctival sac of the rabbit caused mild irritation after a 24-hour exposure period (no further details; Marhold, 1986).

In a further eye irritation study, 50 mg 7-amino-4-hydroxy-2-naphthalenesulfonic acid was instilled into the conjunctival sacs of 2 rabbits. The observation period was 7 days. No irritation was observed (Bayer, 1979).

7.4 Sensitization

No information available.

7.5 Subchronic and chronic toxicity

No information available.

7.6 Genotoxicity

7.6.1 In vitro

7-Amino-4-hydroxy-2-naphthalenesulfonic acid was tested for mutagenicity in the *Salmonella typhimurium* strains TA 98, TA 100, TA 1535, TA 1537 and TA 1538. The study was carried out with the Ames protocol as a direct plate test with and without metabolic activation (S9-mix from Aroclor 1254-induced rat liver) and in a pre-incubation test with metabolic activation in accordance with the method of Prival and Mitchell (1982). With metabolic activation, 5.2 and 41 revertants/ µmol test substance were recorded in the direct plate test and the Prival-modification, respectively (no data on control values given). The mutagenic effect was evaluated as weak and was attributed by the authors to possible contaminants in the product used. Details of the concentrations employed and the strains in which increased revertant counts were seen were not provided in the publication (Freeman et al., 1987).

In a further Salmonella/microsome test, high purity 7-amino-4-hydroxy-2-naphthalenesulfonic acid (I-acid 96.1%, H_2O 4.4%) was tested in the *Salmonella typhimurium* strains TA 98, TA 100, TA 1535 and TA 1537 at concentrations ranging from 8 to

5000 µg/plate, with and without metabolic activation (S9-mix from Aroclor 1254-induced livers from male Sprague-Dawley rats). Concentrations up to and including 40 µg/plate were not bacteriotoxic. At the higher concentrations (from 200 µg/plate), the substance had a weak, strain-specific cytotoxic effect. No biologically relevant increase in the revertant count was observed either with or without metabolic activation. High purity 7-amino-4-hydroxy-2-naphthalenesulfonic acid was therefore not mutagenic in this test system (Bayer, 1992).

7.6.2 In vivo

No information available.

7.7 Carcinogenicity

No information available.

7.8 Reproductive toxicity

No information available.

7.9 Effects on the immune system

No information available.

7.10 Neurotoxicity

No information available.

7.11 Other effects

No information available.

8. Experience in humans

No information available.

9. Threshold limit values

No information available.

References

Bayer AG, Institut für Toxikologie
Akute orale Toxizität von I-Säure an männlichen Wistar-II-Ratten
Unpublished report (1978)

Bayer AG, Institut für Toxikologie
Untersuchung zur Haut- und Schleimhautverträglichkeit von I-Säure an Weißen Neuseeland-Kaninchen
Unpublished report (1979)

Bayer AG, Geschäftsbereich Organica
DIN-Safety data sheet I-Säure (2-Amino-5-naphthol-7-sulfonsäure) (1989)

Bayer AG
Grunddatensatz für Altstoffe über 1000 jato – 2-Naphthalenesulfonic acid, 7-amino-4-hydroxy- (1990)

Bayer AG, Fachbereich Toxikologie
I acid - Salmonella/microsome test
Unpublished report no. 21123 (1992)

Behre, H.
Amino-hydroxynaphthalin-sulfonsäuren
In: Ullmans Encyklopädie der technischen Chemie
4th ed., vol. 17, p. 98–103
Verlag Chemie, Weinheim, New York (1979)

Freeman, H.S., Esancy, J.F., Esancy, M.K., Mills, K.P., Whaley, W.M., Dabney, B.J.
An approach to the design of non-mutagenic azo dyes: 1. The identification of non-mutagenic precursors and potential metabolites
Dyes and Pigments, 8, 417–430 (1987)

Marhold, J.V.
Sbornik vysledku toxikologickeho vysetreni latek a pripravku, p. 105
Institut pro vychovu vedoucicn pracovniku chemickeho prumyclu, Prague, Czechoslovakia (1986)
Cited in: RTECS (1993)

Prival, M.J., Mitchell, V.D.
Analysis of a method for testing azo dyes for mutagenic activity in Salmonella typhimurium in the presence of flavin mononucleotide and hamster liver S9
Mutat. Res., 97, 103–116 (1982)

RTECS (Registry of Toxic Effects of Chemical Substances)
2-Naphthalenesulfonic acid, 7-amino-hydroxy-, QK1295500
produced by NIOSH (National Institute for Occupational Safety and Health) (1993)

6-Amino-4-hydroxy-2-naphthalenesulfonic acid

1. Summary and assessment

6-Amino-4-hydroxy-2-naphthalenesulfonic acid is of very low toxicity on acute oral administration (LD_{50} rat oral > 5000 mg/kg body weight).

In rabbits, 6-amino-4-hydroxy-2-naphthalenesulfonic acid is not irritating to the intact skin or the eye.

6-Amino-4-hydroxy-2-naphthalenesulfonic acid of high purity is not mutagenic in the Salmonella/microsome test.

2. Name of substance

2.1	Usual name	6-Amino-4-hydroxy-2-naphthalenesulfonic acid
2.2	IUPAC-name	6-Amino-4-hydroxynaphthalene-2-sulfonic acid
2.3	CAS-No.	90–51–7
2.4	EINECS-No.	202–000–8

3. Synonyms, common and trade names

2-Amino-8-hydroxynaphthalene-6-sulfonic acid
2-Amino-8-naphthol-6-sulfonic acid
6-Amino-4-naphthol-2-sulfonic acid
7-Amino-1-naphthol-3-sulfonic acid

7-Amino-1-oxy-3-sulfonaphthalene
Aminonaphthol sulfonic acid γ
γ-acid
Gamma acid
6-Amino-4-hydroxy-2-naphthalinsulfonsäure

4. Structural and molecular formulae

4.1 Structural formula

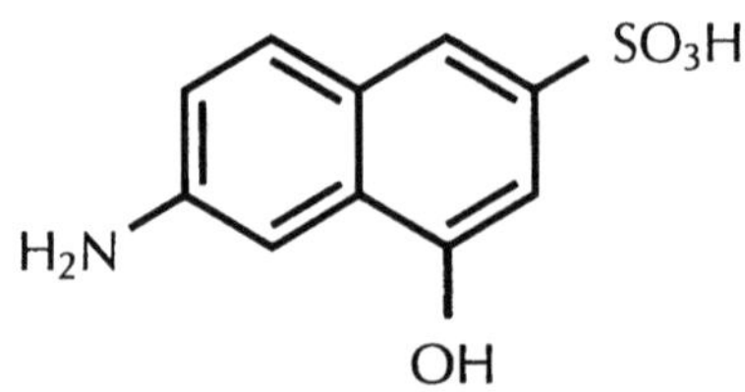

4.2 Molecular formula $C_{10}H_9NO_4S$

5. Physical and chemical properties

5.1	Molecular mass, g/mol	239.25
5.2	Melting point, °C	180–200 (decomposition) (Bayer, 1990)
5.3	Boiling point, °C	No information available
5.4	Vapour pressure, hPa	No information available
5.5	Density, g/cm3	ca. 0.6 (bulk density) (Bayer, 1990)
5.6	Solubility in water	ca. 1 g/l (at 20 °C) (Bayer, 1990) ca. 4 g/l (at 100 °C) (Behre, 1979)
5.7	Solubility in organic solvents	No information available

5.8	Solubility in fat	No information available
5.9	pH value	ca. 3.7 (at 1 g/l water) (Bayer, 1989)
5.10	Conversion factor	1 ml/m^3 (ppm) ≙ 9.77 mg/m^3 1 mg/m^3 ≙ 0.10 ml/m^3 (ppm) (at 1013 hPa and 25 °C)

6. Uses

Coupling component for numerous azo dyes (Behre, 1979).

7. Experimental results

7.1 Toxicokinetics and metabolism

No information available.

7.2 Acute and subacute toxicity

The following oral LD_{50} values have been determined after acute oral exposure:

- for dried, ground gamma acid, minimum content 90%
  rat, male > 5000 mg/kg body weight (Bayer, 1979 a)
- for gamma acid solution (concentration unspecified)
  rat > 15,000 mg/kg body weight (Bayer, 1974 a)

Groups of 10 animals were used per dose. The observation period was 14 days. No signs of toxicity occurred in the rats in either study (Bayer, 1974 a, 1979 a).

7.3 Skin and mucous membrane effects

The application of a solution of 6-amino-4-hydroxy-2-naphthalenesulfonic acid (not further defined) to the skin of 2 rabbits was without effect after an exposure period of 8 h, while after a 24-hour exposure period, mild reddening was observed in 1 of the exposed animals (Bayer, 1974 b).

An aqueous preparation (ca. 500 mg 6-amino-4-hydroxy-2-naphthalenesulfonic acid/animal) was applied to the inner surface of the ears of 2 white rabbits under an adhesive dressing for 24 h and was washed off at the end of the exposure period with soap and water. The observation period was 7 days. No skin irritation was observed (Bayer, 1979 b).

To study the acute eye irritancy, 50 mg 6-amino-4-hydroxy-2-naphthalenesulfonic acid (made into a paste with water) was placed in the conjunctival sacs of 2 rabbits. The observation period was 7 days. Slight reddening, which had cleared up completely after 3 days, was seen in 1 of the 2 exposed rabbits 1 h after application (Bayer, 1979 b).

In a further eye irritation study in rabbits, the instillation of a 6-amino-4-hydroxy-2-naphthalenesulfonic acid solution (no precise details) caused slight reddening for a short time in 2 animals (Bayer, 1974 b).

7.4 Sensitization

No information available.

7.5 Subchronic and chronic toxicity

No information available.

7.6 Genotoxicity

7.6.1 In vitro

6-Amino-4-hydroxy-2-naphthalenesulfonic acid (85% pure) was tested for mutagenicity in the *Salmonella typhimurium* strains TA 98, TA 100, TA 1535 and TA 1537 in the direct plate test with and without metabolic activation (S9-mix from Aroclor 1254-induced rat liver). Concentrations of 0.05 to 5.0 mg/plate were used, and 3 plates were employed per concentration. 6-Amino-4-hydroxy-2-naphthalenesulfonic acid showed strain-specific cytotoxicity at concentrations of 0.5 mg/plate and above. With metabolic activation, weak mutagenic activity was observed in strain TA 98 only, although the revertant count was within the range of historical control values. No mutagenic effect could be observed in the remaining strains either with or without metabolic activation (Mobay, 1984).

In a further study in *Salmonella typhimurium*, 6-amino-4-hydroxy-2-naphthalenesulfonic acid showed a weak mutagenic effect, which in the opinion of the authors was possibly attributable to contaminants in the product tested. The study was carried out in the *Salmonella typhimurium* strains TA 98, TA 100, TA 1535, TA 1537 and TA 1538, with the Ames protocol with and without metabolic activation (S9-mix from Aroclor 1254-induced rat liver), and in a special pre-incubation test with metabolic activation (including 3-fold amounts of S9-mix from non-induced Syrian hamster liver) according to the method of Prival and Mitchell (1982). With metabolic activation, 15.0 and 32.6 revertants/ µmol test substance were recorded in the direct plate test and the Prival-modification, respectively. Details of the concentrations employed and the strains in which increased revertant counts were seen were not given in the publication (Freeman et al., 1987).

High purity 6-amino-4-hydroxy-2-naphthalenesulfonic acid (gamma acid 95%, H_2O 5.9%) was studied in a further Salmonella/microsome test in the *Salmonella typhimurium* strains TA 98, TA 100, TA 1535 and TA 1537 at concentrations ranging from 8 to 5000 µg/plate, with and without metabolic activation (S9-mix from Aroclor 1254-induced livers from male Sprague-Dawley rats). Concentrations up to and including 200 µg/plate showed no bacteriotoxic effects. At the higher concentrations the substance had a weak, strain-specific cytotoxic effect. No biologically relevant increase in the revertant count was observed either with or without metabolic activation. Pure 6-amino-4-hydroxy-2-naphthalenesulfonic acid was therefore not mutagenic under these test conditions (Bayer, 1992).

7.6.2 In vivo

No information available.

7.7 Carcinogenicity

No information available.

7.8 Reproductive toxicity

No information available.

7.9 Effects on the immune system

No information available.

7.10 Neurotoxicity

No information available.

7.11 Other effects

No information available.

8. Experience in humans

No information available.

9. Threshold limit values

No information available.

References

Bayer AG, Institut für Toxikologie
Akute Toxizität von Gamma-Säure Lösung an weiblichen Wistar-Ratten
Unpublished report (1974 a)

Bayer AG, Institut für Toxikologie
Hautfunktionsprüfungen von Gamma-Säure Lösung an Kaninchen
Unpublished report (1974 b)

Bayer AG, Institut für Toxikologie
Akute orale Toxizität von Gammasäure tr. und gem. an männlichen Wistar-II-Ratten
Unpublished report (1979 a)

Bayer AG, Institut für Toxikologie
Untersuchung zur Haut- und Schleimhautverträglichkeit von Gammasäure tr. u. gem. an weißen Neuseeland-Kaninchen
Unpublished report (1979 b)

Bayer AG, Geschäftsbereich Organica
DIN-Safety data sheet Gammasäure (2-Amino-8-naphthol-6-sulfonsäure)(1989)

Bayer AG
Grunddatensatz für Altstoffe über 1000 jato – 2-Naphthalenesulfonic acid, 6-amino-4-hydroxy- (1990)

Bayer AG, Fachbereich Toxikologie
Gamma acid - Salmonella/microsome test
Unpublished report no. 21439 (1992)

Behre, H.
Amino-hydroxynaphthalin-sulfonsäuren
In: Ullmanns Encyklopädie der technischen Chemie
4th ed., vol.17, p.98–103
Verlag Chemie, Weinheim, New York (1979)

Freeman, H.S., Esancy, J.F., Esancy, M.K., Mills, K.P., Whaley, W.M., Dabney, B.J.
An approach to the design of non-mutagenic azo dyes: 1. The identification of non-mutagenic precursors and potential metabolites
Dyes and Pigments, 8, 417–430 (1987)

Mobay Chemical Corporation, Environmental Health Research
The Salmonella/microsome test to evaluate the potential for gamma acid to cause point mutation, study no. 84–500–01
Unpublished report no. 498 (1984)

Prival, M.J., Mitchell, V.D.
Analysis of a method for testing azo dyes for mutagenic activity in *Salmonella typhimurium* in the presence of flavin mononucleotide and hamster liver S9
Mutat. Res., 97, 103–116 (1982)

Tetramethylthiuram monosulfide

1. Summary and assessment

Tetramethylthiuram monosulfide is of moderate toxicity based on the available acute oral studies in animals (LD_{50} rat, oral between 400 and 1390 mg/kg body weight). The following signs of toxicity have been observed in acute studies involving various routes of administration: tachycardia, tachypnoea, miosis, reduced movement, sedation, spasms and paralysis of the extremities. Histologically, signs of hypertrophy of the liver and kidneys have been reported.

In rabbits, tetramethylthiuram monosulfide is slightly irritating to the skin but not irritating to the eyes.

Tetramethylthiuram monosulfide is a skin sensitizer in guinea-pigs.

Long-term oral administration to rats (26 mg/kg body weight/day, 5 days/week for 4 weeks) leads to inhibition of body weight gain and hypertrophy of the hepatocytes and renal tubule cells, as well as reductions in the erythrocyte count and haemoglobin content in the blood.

Tetramethylthiuram monosulfide has been shown to be mutagenic in studies in *Salmonella typhimurium*. The effect is more pronounced with metabolic activation. In eukaryotes, however, tetramethylthiuram monosulfide is not mutagenic. There have also been no indications of an increase in sex-linked recessive lethal mutations in *Drosophila melanogaster*. The remaining mutagenicity studies carried out on this substance (sister chromatid exchange rate in vitro and micronucleus test in the hamster) do not permit a conclusive evaluation.

The available results of studies on the carcinogenicity of tetramethylthiuram monosulfide in the mouse after oral and subcutaneous administration give no indications of a treatment-related increase in the tumour rate. However, the available data do not permit a conclusive evaluation, due to the small number of animals used. The results of a cell transformation test were also negative.

There were no indications of teratogenicity in the mouse after subcutaneous administration in a screening test.

Tetramethylthiuram monosulfide inhibits microsomal monooxygenases and cytosolic enzyme systems, particularly alcohol and aldehyde dehydrogenase, both in vitro and in vivo.

It is likely that tetramethylthiuram monosulfide may cause alcohol intolerance reactions in man due to its ability to inhibit alcohol and aldehyde dehydrogenase activity. The substance can cause sensitization and contact dermatitis in man.

2. Name of substance

2.1	Usual name	Tetramethylthiuram monosulfide
2.2	IUPAC-name	Bis(dimethylthiocarbamoyl)sulfide
2.3	CAS-No.	97–74–5
2.4	EINECS-No.	202–605–7

3. Synonyms, common and trade names

Bis(dimethylamino)thioxomethyl sulfide
Bis(dimethylthiocarbamyl)-monosulfide
Dimethylcarbamothioic acid anhydrosulfide
Dimethyldithiocarbamic acid anhydrosulfide
Ekagom TM
Monex
Mono Thiurad
Monothiurad
Monothiuram

Nocceler TS
Pennac MS
Tetramethylthiocarbamic acid anhydrosulfide
Tetramethylthiodicarbonic diamide
Tetramethylthiuram sulfide
Tetramethylthiurammono sulfid
Tetramethylthiurammonium sulfide
N,N,N′,N′-Tetramethylthiuram monosulfide
Tetramethyltrithio-carbamic anhydride
1,1′-Thiobis(N,N-dimethylthio)formamide
Thionex
Thionex-rubber accelerator
Thiuram MM
TMTM
TMTMS
UNADS
USAF B-32
USAF EK-P-6255
Vulkacit Thiuram MS/C

4. Structural and molecular formulae

4.1 Structural formula

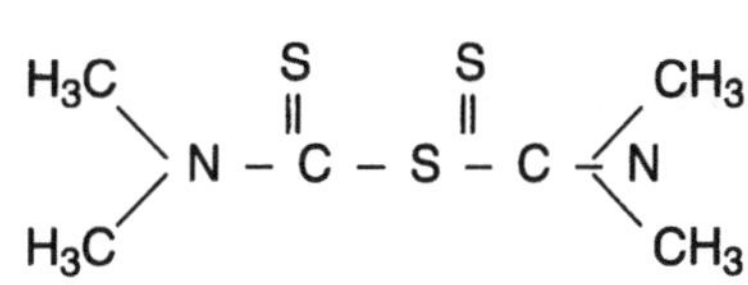

4.2 Molecular formula $C_6H_{12}N_2S_3$

5. Physical and chemical properties

5.1	Molecular mass, g/mol	208.38
5.2	Melting point, °C	104 (Torres Alanis et al., 1982) 107 (Gaul, 1957)
5.3	Boiling point, °C	No information available
5.4	Vapour pressure, hPa	No information available
5.5	Density, g/cm3	1.39 (at 20 °C) (Gaul, 1957)
5.6	Solubility in water	Insoluble (at 20 °C) (Gaul, 1957)
5.7	Solubility in organic solvents	Soluble in acetone, benzene and chloroform; insoluble in petrol (Gaul, 1957)
5.8	Solubility in fat	No information available
5.9	pH value	No information available
5.10	Conversion factor	1 ml/m^3 (ppm) ≙ 8.51 mg/m^3 1 mg/m^3 ≙ 0.12 ml/m^3 (ppm) (at 1013 hPa and 25 °C)

6. Uses

Vulcanization accelerator in the rubber industry (Schubart, 1987).

7. Experimental results

7.1 Toxicokinetics and metabolism

No information available.

7.2 Acute and subacute toxicity

The following LD_{50} values have been determined on acute oral and intraperitoneal administration of tetramethylthiuram monosulfide:

Mouse	oral	between 600 and 7400 mg/kg body weight
Mouse	intraperitoneal	between 80 and 565 mg/kg body weight
Rat	oral	between 400 and 1390 mg/kg body weight
Rat	intraperitoneal	383 mg/kg body weight

For further details see Table 1.

The following signs of toxicity have been reported in rats and mice after oral or intraperitoneal administration of tetramethylthiuram monosulfide during acute toxicity studies: excitation, tachycardia, tachypnoea, miosis, reduced movement, sedation, spasms and paralysis of the extremities, narcosis, diarrhoea, weight loss, rough coats and hair loss (Korablev, 1965; Poitou et al., 1978; Torres Alanis et al., 1982; Bayer, 1980).

Histopathological examination of the mouse following oral administration of 867 mg tetramethylthiuram monosulfide/kg body weight revealed hypertrophy of the hepatocytes, Kupffer's cells and kidney tubule cells (Torres Alanis et al., 1982).

The oral administration of tetramethylthiuram monosulfide to rats at a dose of 1/5 of the LD_{50} (no further details) caused a reduction in diuresis of ca. 75% and in saluresis of 70 to 80% within 4 h (Korablev and Evets, 1977).

The minimum lethal doses (MLD) for tetramethylthiuram monosulfide after intraperitoneal injection were 1 and 5 mg/kg body weight in mice and rats, respectively. After oral administration, the values were 10 mg/kg body weight in marmots and 100 mg/kg body weight in rabbits and dogs (no further details; Takahashi, 1975).

Male rats were exposed to tetramethylthiuram monosulfide at a concentration of 400 mg/m^3 2 h/day for 14 days. A reduction in body weight and an increase in chronaxy were observed. The protein content of the urine increased from 0.77 to 1.39% in the exposed animals, while the amino nitrogen concentration increased

Table 1. Acute toxicity of tetramethylthiuram monosulfide

Species	Number of animals/dose	Sex	Route	Observation period (days)	no effect level (mg/kg bw)	LD_{50} (95% confidence limits) (mg/kg bw)	Reference
Mouse	–	–	oral	–	–	600	Mitani, 1960
Mouse	10–20	female	oral	1	< 500	818 (583–995)	Torres Alanis et al., 1982
Mouse	–	–	oral	–	–	1150 (951.5–1348.5)	Vorob'eva, 1964, 1965
Mouse	–	–	oral	–	–	1200 (1010–1390)	Korablev, 1965
Mouse	–	–	oral	–	–	2950 (2366–3523)	Korablev, 1969
Mouse	–	–	oral	–	–	7400 (5700–9100)	Korablev, 1969
Mouse	–	–	i.p.	–	< 1	–	McCormick, 1971
Mouse	–	–	i.p.	–	–	ca. 80	Korablev, 1969
Mouse	> 10	–	i.p.	–	–	250	Hald et al., 1952
Mouse	–	–	i.p.	–	–	250	Korablev, 1969
Mouse	–	male	i.p.	7	–	300–500	Doull et al., 1962
Mouse	20	10 male 10 female	i.p.	14	–	565 (520–610)	Poitou et al., 1978
Mouse	–	–	i.p.	–	–	800	Takahashi, 1975
Rat	–	–	oral	–	–	400 (300–500)	Korablev, 1965
Rat	2 animals in total	–	oral or i.p.	7	–	> 500	Dieke et al., 1947

Table 1. Acute toxicity of tetramethylthiuram monosulfide

Species	Number of animals/dose	Sex	Route	Observation period (days)	no effect level (mg/kg bw)	LD_{50} (95% confidence limits) (mg/kg bw)	Reference
Rat	10	male	oral	14	< 100	690	Bayer, 1980
Rat	–	–	oral	–	–	900	Davydov et al., 1980
Rat	–	–	oral	–	–	1390	Schunk, 1984
Rat	–	–	i.p.	–	< 5	–	McCormick, 1971
Rat	20	10 male 10 female	i.p.	14	–	383 (361–403)	Poitou et al., 1978
Guinea pig	–	–	oral	–	< 10	–	McCormick, 1971
Guinea pig	–	–	i.p.	–	< 10	–	McCormick, 1971
Rabbit	–	–	oral	–	< 100	–	McCormick, 1971
Cat	–	–	oral	–	< 100	–	McCormick, 1971
Dog	–	–	oral	–	< 100	–	McCormick, 1971

bw body weight
– no data

from 0.75 to 3.4 mg% (mg/100 ml). Paper chromatography revealed the presence of histidine, methionine, glutamic acid, aspartic acid, cystine, arginine, lysine, threonine, tryptophan and phenylalanine in the urine of the exposed animals, while in the urine of the controls, histidine, methionine, glutamic acid, aspartic acid and glycine were detectable. The SH-content was slightly reduced following treatment (no further details; Vorob'eva, 1965).

The oral administration of 26 mg tetramethylthiuram monosulfide/kg body weight/day to female Wistar rats 5 days/week for 4 weeks by gavage, led to a reduction in body weight gain of about 10% compared with the controls. Each group contained 10 females. Feed consumption was reduced by ca. 20 to 30% and water intake was increased by about 40%. The relative liver weights were significantly increased ($p < 0.01$). The activities of glucose-6-phosphatase and O-demethylase and the levels of phosphatidyl choline, malondialdehyde, cytochrome P-450 and protein in the liver were unaffected. The erythrocyte count and the haemoglobin content were significantly reduced ($p \leq 0.01$) in all treated animals. The leukocyte count, the levels of protein, urea and creatinine and the activities of alkaline phosphatase and alanine aminotransferase were unaffected. Histological examination revealed hypertrophy of the hepatocytes. Slight hypertrophy of the kidney tubule cells was observed in all rats in the treated group. There were no indications of histopathological effects in the lungs, heart, spleen, muscle tissue, brain or ischiatic nerve (Torres Alanis et al., 1982).

7.3 Skin and mucous membrane effects

The hair was removed from both flanks of 6 male albino rabbits (New Zealand white, weight ca. 2.9 kg) 24 h before the beginning of a skin irritation study; the right flank was also scarified. The substance was applied (500 mg mixed to a paste with water) for 24 h under occlusive cover. The effects were recorded 24, 48 and 72 h and 8 days after the beginning of the study. The intact skin of one animal showed slight reddening after 24 h, while that of all of the animals showed slight oedema. The scarified skin also showed mild symptoms of skin irritation. The substance was therefore evaluated as slightly irritating (FhG, 1980 a).

The possible eye irritancy of the substance was studied in 6 male albino rabbits (New Zealand white, weight ca. 2.9 kg). A single dose of 100 µl (apparent volume) of the amorphous substance was placed in the conjunctival sac of the lower right eyelid. The untreated left eye served as a control. The results were read after 24, 48 and 72 h and after 8 days. Slight reddening of the conjunctiva, which had cleared up after 8 days, was seen in 4 animals. Based on these results, the substance was evaluated as not irritating to the eye (FhG, 1980 b).

In a safety data sheet, tetramethylthiuram monosulfide was evaluated as practically non-irritating to the skin (irritation index 0.7 from a maximum of 8.0) and slightly irritating to the eye (irritation index 4.0 from a maximum of 110.0; no further details; Monsanto, undated).

7.4 Sensitization

The skin sensitizing properties of tetramethylthiuram monosulfide were tested in guinea-pigs in an occlusive epicutaneous test. A single application of a 0.5 M tetramethylthiuram monosulfide suspension in vaseline (equivalent to 104 mg/ml) was used for induction, while the challenge treatment consisted of 0.05 M (equivalent to 10.4 mg/ml) or 0.5 M suspensions. The skin reactions were evaluated 72 h after the beginning of the challenge treatment. Tetramethylthiuram monosulfide was sensitizing in 3 of 12 animals in the 0.05 M group and 9 of 12 animals in the 0.5 M group. It also caused a cross-reaction in guinea pigs induced with dimethylamine. The treatments used for induction and challenge were 0.5 M dimethylamine and 0.5 M tetramethylthiuram monosulfide suspensions, respectively. No cross-reactions were observed with 2-(N',N-diethylthiocarbamoylthio)benzothiazole or morpholinothiobenzothiazole (Kantoh et al., 1985).

7.5 Subchronic and chronic toxicity

Male rabbits were given 20 mg tetramethylthiuram monosulfide/kg body weight/day orally every other day for 2 months, and then daily for a further 2 months. Body weight gain was slightly retarded (minus 10%). The haemoglobin and albumin content of the serum, and the thromboplastin time were reduced by about 15%, the erythrocyte count by 30%, the leukocyte count by about 73% and the alkaline phosphatase activity by ca. 82%. The amino acid content in the serum was increased by 30%. Macroscopic effects seen in the animals killed at the end of the study and in those that died prematurely included bleeding in the liver, heart muscle and mesenterium and pneumonia. The relative lung weight was increased by ca. 46%. Histological examination revealed pulmonary oedema and marked "dystrophy" of the liver and kidneys (no further details; Vorob'eva, 1964, 1965).

7.6 Genotoxicity

7.6.1 In vitro

Tetramethylthiuram monosulfide was tested for mutagenic activity in prokaryotes (*Salmonella typhimurium* strains TA 98, TA 100, TA 1535, TA 1537 and TA 1538) in the direct plate test at concentrations of from 5 to 2000 µg/plate. The tetramethylthiuram monosulfide lots used were of analytical grade purity and were recrystallized. For each concentration and strain, 3 to 6 plates were used. The studies were carried out with and without metabolic activation (S9-mix from Aroclor 1254-induced rat liver). Concentrations of 1000 µg/plate and above were in the cytotoxic range. Tetramethylthiuram monosulfide was mutagenic in strains TA 100 and TA 1535 both with and without metabolic activation. Without metabolic activation, a concentration-dependent increase in the revertant count (maximum 2.5-fold compared with controls) was seen in both strains between 10 and 1000 µg/plate; on addition of S9-mix, the revertant count was increased by a maximum of about 5-fold. In strain TA 1537, tetramethylthiuram monosulfide caused a marginal increase in the revertant count (maximum 2-fold) with and without metabolic activation at concentrations of between 500 and 2000 µg/plate. There were no indications of mutagenic activity in strains TA 98 and TA 1538 either with or without metabolic activation (Hedenstedt et al., 1979; Hedenstedt, 1981; Torres Alanis et al., 1982).

In a further Ames test in the *Salmonella typhimurium* strains TA 98, TA 100, TA 1535 and TA 1537 without metabolic activation, a significant dose-dependent increase in revertants/plate was seen in strain TA 100, and an increase that was not dose-related was seen in TA 1537 at the tested concentrations of from 10 to 320 µg/plate. Tetramethylthiuram monosulfide was therefore evaluated as being mutagenic (Goodyear, 1980).

Concentration-dependent increases in the revertant count (maximum 3.6- and 2.5-fold) were observed in the *Salmonella typhimurium* strains TA 98 and TA 100, respectively at concentrations of from 50 to 1000 µg tetramethylthiuram monosulfide/plate with the addition of S9-mix (from PCB-induced rat liver). No indications of mutagenic activity were seen without S9-mix at concentrations of from 10 to 1000 µg tetramethylthiuram monosulfide/plate. The high concentrations tested were not in the cytotoxic range (You et al., 1982).

In a further study, tetramethylthiuram monosulfide was evaluated as mutagenic in *Salmonella typhimurium* from 5 µg/plate (no further details; Rannug et al., 1984).

Tetramethylthiuram monosulfide was tested at 6 concentrations of between 10 and 1000 µg/plate in a plate incorporation test in the *Salmonella typhimurium* strains TA 97, TA 98, TA 100 and TA 102. The substance was positive in strains TA 97 and TA 100 without metabolic activation, and in all strains with metabolic activation. In a second experiment with pre-incubation, the test substance was similarly mutagenic in all strains with metabolic activation. At the top dose, tetramethylthiuram monosulfide caused toxicity without metabolic activation in strains TA 97, TA 98 and TA 102, in the form of microcolonies, reduced bacterial lawns and a fall in the number of revertant colonies (Uniroyal, 1984).

Tetramethylthiuram monosulfide was tested at concentrations of from 0.1 to 500 µg/plate in the *Salmonella typhimurium* strains TA 98, TA 100, TA 1535, TA 1537 and TA 1538 and in *Saccharomyces cerevisiae* D4 with and without metabolic activation (S9-mix from Aroclor 1254-induced livers from male Sprague-Dawley rats). The substance was negative without metabolic activation. With metabolic activation there was an increase in the revertant count in strain TA 1535 from 10 µg/plate (LBI, 1976).

Technical grade tetramethylthiuram monosulfide was not mutagenic after exposure for 4 h either with or without metabolic activation (S9-mix from Aroclor 1254-induced rat liver) in the HPRT test in

Chinese hamster V79 cells. The concentrations tested were between 0.2 and 3.2 µg/ml medium with the high concentrations being cytotoxic (Donner et al., 1983).

Tetramethylthiuram monosulfide was negative in tests in L5178Y mouse lymphoma cells both with and without metabolic activation (S9-mix from the livers of male HA/ICR mice; no details of induction) at concentrations of from 0.02 to 40 µg/ml. Concentrations of > 10 µg/ml were cytotoxic (LBI, 1978).

The effect of technical grade tetramethylthiuram monosulfide on the sister chromatid exchange rate was studied in Chinese hamster cells (CHO cell line). Concentrations of from 10^{-7} to 2×10^{-5} M (0.2 to 4.17 µg/ml) were tested. Tetramethylthiuram monosulfide was cytotoxic at the high concentration. The study was carried out with and without metabolic activation (S9-mix from Clophen-induced rat liver). Between 57 and 90 metaphases were evaluated per concentration, with 296 and 326 evaluated in the two control groups. Without metabolic activation, tetramethylthiuram monosulfide led to a slight but significant ($p < 0.01$, student's t-test) increase in the exchange rate at the low and intermediate concentrations (10.23 and 10.89 exchanges/metaphase) compared with the controls (9.6). The exchange rate at the second highest (highest evaluatable) concentration was within the range of that in the control group. With metabolic activation, the sister chromatid exchange rate did not differ between the treated and control rats. Because tetramethylthiuram monosulfide did not double the sister chromatid exchange rate in this study, it was not considered by the authors to be a potent inducer of sister chromatid exchange (Donner et al., 1983).

7.6.2 In vivo

In the micronucleus test, pairs of male Chinese hamsters were given single intraperitoneal injections of 250 and 500 mg tetramethylthiuram monosulfide/kg body weight, and 2 males and 3 females were given 100 mg/kg body weight. A control group (4 males and 6 females) were treated with the carrier, olive oil. Bone marrow preparations were made 30 h after administration. About 2000 polychromatic and 2000 normochromatic erythrocytes from each animal were examined for the presence of micronuclei. All doses reduced the ratio of polychromatic to normochromatic erythrocytes by about 2- to 6-fold, indicating cytotoxicity. No dose-effect relationship was evident. An increase in the occurrence of micronuclei in the polychromatic erythrocytes (maximum ca. 2-fold) was seen at 250 and 500 mg tetramethylthiuram monosulfide/kg body weight, although the increase was only significant ($p \leq 0.05$) at 250 mg/kg body weight. The occurrence of micronuclei in normochromatic erythrocytes was unchanged. According to the authors, the substance was mutagenic in this test system (Donner et al., 1983).

Technical grade tetramethylthiuram monosulfide caused no increase in sex-linked recessive lethal mutations in *Drosophila melanogaster* at concentrations of from 0.04 to 4.0 mg/ml nutrient solution. The high concentration was within the maximum tolerable range (no further details; Donner et al., 1983).

No increase in sex-linked recessive lethal mutations was observed in an earlier study by the same group (no further details; Donner, 1981).

A further study in Drosophila showed that a concentration of 750 µg tetramethylthiuram monosulfide/ml in the nutrient solution (4% sucrose and 0.5% Tween 80 in distilled water) did not significantly increase the number of recessive lethal mutations in the treated group (0.14%) compared with the controls (0.11%; Goodyear, 1982).

7.7 Carcinogenicity

Groups of 18 male and 18 female B6C3F1 and B6AKF1 mice were given tetramethylthiuram monosulfide by gavage (in 0.5% gelatine) from day 7 to day 28 of life. The dose was equivalent to 100 mg/kg body weight at the beginning of the study, but was not adjusted to body weight gain. From day 28 onwards, the mice received tetramethylthiuram monosulfide in the diet at a concentration of 377 ppm (equivalent to 100 mg/kg body weight, based on the average body weight of 15 g on day 28). The dose was equivalent to the maximum tolerated dose determined in a preliminary 19-day dietary study in 6 mice. The duration of exposure was 18 months. Control groups consisted of 18 male and 18 female B6C3F1 and B6AKF1 mice, which received the carrier substance. The animals that died prematurely and those killed at the end of the study were examined macroscopically for tumours and lesions. Haematological evaluation of blood smears was carried out in mice with enlargement of the spleen or liver or with lymphadenopathy. Histological examination of the liver, spleen, kidneys, adrenal glands, stomach, gut, genital organs and brain were carried out in all of the animals sacrificed at the end of the study. In the animals that died during the study, histological examination was carried out on a case-by-case basis (no further details). No effects on body weight were found when the mice were weighed after 1, 6, 12 and 20 months. No data on feed consumption were provided in the study report (Bionetics, 1968 a). Mortality, tumour incidences, types and locations of tumours and non-neoplastic effects are shown in Tables 2 and 3:

Table 2. Carcinogenicity study in B6AKF1 mice involving oral administration for 18 months

	Controls		Tetramethylthiuram monosulfide*	
	Male	Female	Male	Female
Mice exposed	18	18	18	18
Mice dying prematurely	0	3	1	1
Mice examined	18	17	17	17
Number of mice without tumours	15	15	16	14
Number of mice with tumours	3	2	1	3
% of mice with tumours	16.7	11.8	5.6	22.2
<u>Type of tumour</u>				
Lymphatic leukaemia	0	1	0	0
Reticulum cell sarcoma	0	0	0	2
Lung adenoma	2	1	1	0
Lung carcinoma	0	0	0	0
Hepatoma	1	0	0	1
Mammary carcinoma	0	1	0	1
<u>Lesions</u>				
Focal pneumonia	0	0	2	2
Aspiration pneumonia	2	3	0	0
Perirenal infiltration	0	1	0	0
Hperplasia of Peyer's patches	0	1	0	0
Lymphoid lung infiltration	0	1	0	0

* see text for dose

Table 3. Carcinogenicity study in B6C3F1 mice involving oral administration for 18 months

	Controls		Tetramethylthiuram monosulfide*	
	Male	Female	Male	Female
Mice exposed	18	18	18	18
Mice dying prematurely	2	2	1	2
Mice examined	16	16	17	16
Number of mice without tumours	16	16	12	16
Number of mice with tumours	0	0	5	0
% of mice with tumours	0	0	29.4	0
Type of tumour				
Lymphatic leukaemia	0	0	0	0
Reticulum cell sarcoma	0	0	0	0
Lung adenoma	0	0	3	0
Lung carcinoma	0	0	0	0
Hepatoma	0	0	2	0
Mammary carcinoma	0	0	0	0
Spleen angioma	0	0	1	0
Stomach papilloma	3	4	0	0
Lesions				
Focal pneumonia	0	0	5	2
Follicular hyperplasia	3***	0	2**	3**
Lympoid infiltration	1	0	2	1
Focal gastritis/hyperkeratosis of the stomach	0	0	2	0
Subcutaneous abscess	0	0	1	0
Aspiration pneumonia	3	4	0	0
Liver necrosis	1	0	0	0
Haemorrhagic nodulus	1	0	0	0

* see text for dose ** no detailed information provided *** spleen

In a further carcinogenicity study, groups of 18 male and 18 female B6C3F1 and B6AKF1 mice were injected subcutaneously (in the neck) with 100 mg tetramethylthiuram monosulfide/kg body weight on day 28 of life. The carrier, dimethyl sulfoxide, was administered to a further 24 animals per sex and strain. The observation period was 18 months. Mice killed at the end of the study and those that died prematurely were examined for tumours and lesions. Haematological evaluation of blood smears was carried out in mice with enlargement of the spleen or liver or with lymphadenopathy. Histological examination of the liver, spleen, kidneys, adrenal glands, stomach, gut, genital organs and brain were carried out in all of the animals sacrificed at the end of the study. In the animals that died during the study, histological examination was carried out on a case-by-case basis (no further details; Bionetics, 1968 a). The results of the histopathological evaluation are shown in Tables 4 and 5.

Table 4. Carcinogenicity study in B6AKF1 mice involving a single subcutaneous injection (18-month observation period)

	Controls		Tetramethylthiuram monosulfide*	
	Male	Female	Male	Female
Mice exposed	24	24	18	18
Mice dying prematurely	2	0	0	0
Mice examined	24	24	18	18
Number of mice without tumours	23	21	14	16
Number of mice with tumours	1	3	4	2
% of mice with tumours	4.1	12.5	22.2	11.1
Type of tumour				
Lymphatic leukaemia	0	0	0	0
Reticulum cell sarcoma	0	1	0	0
Lung adenoma	1	2	4	1
Lung carcinoma	0	0	0	0
Hepatoma	0	0	0	0
Mammary carcinoma	0	0	0	0
Uterine leiomyoma	–	0	–	1
Lesions				
Focal pneumonia	3	4	0	0
Lymphoid infiltration	0	2	0	0
Lymphoid liver nodulus	0	1	0	0
Mesenterial granuloma	1	0	0	0

Table 5. Carcinogenicity study in B6C3F1 mice involving a single subcutaneous injection (18-month observation period)

	Controls		Tetramethylthiuram monosulfide*	
	Male	Female	Male	Female
Mice exposed	24	24	18	18
Mice dying prematurely	1	1	3	1
Mice examined	24	23	18	17
Number of mice without tumours	22	22	15	17
Number of mice with tumours	2	1	3	0
% of mice with tumours	8.3	4.3	16.7	0
Type of tumour				
Lymphatic leukaemia	0	0	0	0
Reticulum cell sarcoma	0	0	2	0
Lung adenoma	0	1	0	0
Lung carcinoma	0	0	0	0
Hepatoma	1	0	1	0
Mammary carcinoma	0	0	0	0
Spleen angioma	1	0	0	0
Lesions				
Nephritis with fibrosis	0	0	1	0
Focal pneumonia	4	2	0	0
Seminal vesicle cyst	1	-	0	-
Lymphoid atrophy of the spleen	1	0	0	0
Lymphoid infiltration	0	1	0	0

The available studies on the carcinogenicity of tetramethylthiuram monosulfide do not permit a conclusive evaluation of its carcinogenic potential, for the following reasons, among others:

- the number of animals was relatively low (maximum 24; current requirements are 50 males/50 females and 100 controls of each sex)
- no exact details were given of the doses administered or whether the doses were within the maximum tolerable range
- no data were presented on the tumour incidence in historical controls in the chosen strains.

The available data give no clear indication of tetramethylthiuram monosulfide being carcinogenic.

Concentrations of 0.06 to 1000 µg tetramethylthiuram monosulfide/ml were studied in a cell transformation test in BALB/3T3 cells. No surviving cells were observed at concentrations of 0.98 µg/ml and above. Concentrations of from 0.038 to 0.45 µg/ml were therefore used in the transformation test. None of these concentrations increased the number of transformed foci (0.13 to 0.27) above the spontaneous value (0.33; LBI, 1981).

7.8 Reproductive toxicity

BL6 mice were injected subcutaneously in the neck with 46.4 or 100 mg tetramethylthiuram monosulfide/kg body weight/day from day 6 to day 14 of pregnancy and AKR mice were injected with 100 mg tetramethylthiuram monosulfide/kg body weight/day from day 6 to day 15 of pregnancy. Controls were treated in the same manner with the carrier, dimethyl sulfoxide (DMSO). Post-mortems were carried out on day 18 of pregnancy in BL6 mice and on day 19 of pregnancy in AKR mice. In each of the treated groups, 6 litters were examined with the exception of the 100 mg/kg group of BL6 mice (1 litter) and the control groups (73 and 36 in litters BL6 and AKR mice, respectively). In the AKR mice at 100 mg/kg, maternal body weight was significantly reduced ($p < 0.05$), while relative liver weights were within the range of the DMSO control values. Placental weight was clearly increased, the volume of amniotic fluid per foetus markedly reduced. The numbers of implants per litter and live foetuses corresponded with the control values. Foetal mortality was 9% (controls 16%), while foetal weight was significantly below that in the controls. No foetal malformations occurred (41 foetuses examined). In the BL6 mice at 46.4 mg/kg, the relative liver weight was significantly increased in the dams ($p < 0.05$), while body and placental weights and the volume of amniotic fluid per foetus corresponded to control values. The numbers of implants per litter and live foetuses and the foetal mortality were within normal limits. Foetal weight was significantly increased. Malformations occurred in 12% of the foetuses (controls 9%; 41 foetuses examined). Anophthalmia was observed in 2 foetuses and microphthalmia in 4, malformations that were also seen in the control group. In the one dam in the 100 mg/kg group, body weight, relative liver weight and placental weight corresponded to control values, as did the volume of amniotic fluid per foetus. The numbers of implants and live foetuses, foetal weight and foetal mortality were within the range of control values. One of the 8 foetuses was malformed with micro- and anophthalmia, micrognathia and cleft palate. The authors considered that the substance had not caused any "important" teratogenic effects ("Unads did not produce important effects"; Bionetics, 1968 b).

The effect of tetramethylthiuram monosulfide on post-natal development was studied in 6 spontaneously delivered litters of BL6 mice. The dams were given subcutaneous injections of 46.4 mg tetramethylthiuram monosulfide/kg body weight on days 6 to 14 of pregnancy and the offspring were examined for skeletal and visceral malformations on day 8 of life. The number of live foetuses per litter, live weight on day 1 and mortality up to day 8 were within the range of control values. Anomalies were seen in 6/43 (14%) of the offspring (controls 4/138, 2.9%) including micro- and anophthalmia, exophthalmos and agnathia, malformations which were also seen in the controls. This study was not evaluated statistically because of the low number of experimental animals and no statement can therefore be made on the teratogenicity of tetramethylthiuram monosulfide (Bionetics, 1968 b).

Because of the low numbers of animals used in both cases (Bionetics, 1968b), these studies only give a preliminary insight into the toxic effects of tetramethylthiuram monosulfide on reproduction.

White leghorn chicken eggs were used to study the embryotoxicity (mortality and malformations) of tetramethylthiuram monosulfide. The chemical was injected into the air chamber of hens' eggs containing 3-day-old embryos at doses of 15 to 100 nmol/egg (equivalent to 3 to 20.8 µg/egg). Between 20 and 40 chicken embryos were treated per dose and 70 control embryos were treated with the vehicle, acetone. Dead embryos were picked out every 2 to 3 days. The observation period was 11 days. Parameters used for the evaluation of embryotoxicity included "early" (within 2 days of treatment) and "late" (within 2 to 11 days of treatment) dead malformed and dead not malformed embryos and surviving malformed chicks.

The ED_{50} value (embryotoxicity) was reported as 7.0 µg/egg (95% confidence limits 5.8 to 8.5). Tetramethylthiuram monosulfide led to a concentration-dependent increase in mortality at doses of from 3 µg/egg during the first 2 days after treatment. The number of malformations in the surviving birds was increased to 15% at the high concentration (controls 0%). Malformations included small eye cups, defects of the cornea and eyelid, open coeloms and malformations of the extremities. However, tetramethylthiuram monosulfide caused relatively few malformations at any of the concentrations tested (Korhonen et al., 1982, 1983). The application of these findings to mammalian systems is, however, not really possible, as the absorption barriers and detoxification mechanisms present in mammals are not represented in this system. Furthermore, this test system is over-sensitive and is unable to distinguish between teratogenicity and embryolethal effects (Skofitsch, 1988; Neubert, 1993; Heinrich-Hirsch, 1992).

7.9 Effects on the immune system

In the lymphocyte transformation test (LTT) hapten-protein conjugates were used to introduce allergens into lymphocyte cultures and the proliferation of lymphocytes following incubation with these conjugates was determined. On the assumption that the relative antigenic determinant of tetramethylthiuram monosulfide is the dimethylthiocarbamoyl group, dimethylthiocarbamoyl-HSA (human serum albumin) and dimethylcarbamoyl-HSA conjugates were produced. Both conjugates were incubated for 7 days at 37 °C with lymphocytes from 4 patients who reacted positively to tetramethylthiuram monosulfide in patch tests, 2 patients who reacted negatively to thiuram-mix in patch tests and a control person, at concentrations of 0.5 to 100 µg/ml. Only the lymphocytes from the 4 subjects with positive patch tests reacted to the hapten-protein conjugates with varying degrees of proliferation. The authors considered that this test system could be used as a supplement or an alternative to the patch test in the diagnosis of skin sensitization (Kimber et al., 1991).

7.10 Neurotoxicity

No information available.

7.11 Other effects

Tetramethylthiuram monosulfide has sedative and narcotic effects in rats and mice:

The spontaneous movement of Swiss-albino mice was significantly reduced ($p \leq 0.001$) 30 min after intraperitoneal injection of tetramethylthiuram monosulfide from doses of 100 mg/kg body weight. Tetramethylthiuram monosulfide did not inhibit reflexes in the rotarod test at this dose (Poitou et al., 1978).

The intraperitoneal injection of 40 mg tetramethylthiuram monosulfide/kg body weight caused a significant ($p \leq 0.001$) reduction in spontaneous movement (by ca. 70%) in Wistar rats 2 h after administration. Body temperature was reduced from 37.0 to 34.7 °C after 2 h (Poitou et al., 1978).

The escape behaviour of conditioned male Wistar rats was reduced by 54% by the intraperitoneal injection of 40 mg tetramethylthiuram monosulfide/kg body weight (Poitou et al., 1978).

In mice, the intraperitoneal administration of 80 mg tetramethylthiuram monosulfide/kg body weight significantly ($p \leq 0.001$) inhibited increases in motility induced by L-DOPA (250 mg/kg body weight by intraperitoneal injection) by ca. 70% (Poitou et al., 1978).

The activities of microsomal monooxygenases and cytosolic enzyme systems were inhibited by tetramethylthiuram monosulfide both in vitro and in vivo. The pharmacodynamic effects in animals described in the second part of this section (including increases in the hexobarbital sleeping time and zoxazolamine paralysis) are probably due to inhibition of these enzyme systems. It has been proposed (Kitson, 1976) that tetramethylthiuram monosulfide forms a covalent bond with the sulfur bridges of enzymes or inhibits the incorporation of palmitic acid into the endoplasmic reticulum (Torres Alanis et al., 1982).

The pretreatment of 4 female Wistar rats with an oral dose of 16 µmol tetramethylthiuram monosulfide/kg body weight (equivalent to 3.3 mg/kg body weight) 90 min or 18 h before the administration of ethanol (2 g/kg body weight by intraperitoneal injection) led to a significant ($p \leq 0.01$) increase in the ethanol and acetaldehyde levels in the blood. When tetramethylthiuram monosulfide pretreatment was carried out 90 min before ethanol administration, the acetaldehyde level 4 h after ethanol administration was ca. 7-fold higher than that in the control group (n=28 rats), while with pretreatment 18 h before, it was ca. 6-fold higher. Pretreatment with tetramethylthiuram monosulfide 48 h before ethanol administration doubled the acetaldehyde level 4 h after ethanol administration. The half-life for elimination of ethanol increased from ca. 140 min in the control group to ca. 210 min in rats given tetramethylthiuram monosulfide pretreatment 90 min before ethanol administration. It was suggested that inhibition of alcohol- and/or aldehyde dehydrogenase takes place (Garcia de Torres et al., 1983; Römer et al., 1982, 1984).

Adult albino rabbits (2 animals, weight 2.0 to 3.5 kg) were given a control injection of 1250 mg ethanol/kg body weight. 18 to 20 h before a second dose of ethanol, the animals were given 300 mg tetramethylthiuram monosulfide/kg body weight orally. Within 1 h of the alcohol administration, a 9.6-fold higher acetaldehyde level was detected in the blood. After 3 h, the value was 16.9-fold higher than the control value determined before the administration of tetramethylthiuram monosulfide (Akabane and Ikomi, 1958).

A clearly increased acetaldehyde level was also observed in a further study in rabbits after the oral administration of 300 mg tetramethylthiuram monosulfide/kg body weight followed by intravenous ethanol administration (1000 mg/kg body weight; no further details; Mitani, 1960).

Pretreatment of rabbits with 80 or 140 mg tetramethylthiuram monosulfide/kg body weight orally 1 to 25 h before administration of ethanol (1500 mg/kg body weight) led to a ca. 10-fold increase in the acetaldehyde level. Each group consisted of 1 or 2 animals. No dose- or time-dependency on tetramethylthiuram monosulfide administration could be derived from the available data (Hald et al., 1952).

In vitro, tetramethylthiuram monosulfide inhibited the activities of alcohol dehydrogenase from rat liver cytosol and aldehyde dehydrogenase from mitochondria or endoplasmic reticulum. The alcohol dehydrogenase activity was inhibited by 61% by 1.0 mmol tetramethylthiuram monosulfide/l (equivalent to 208.38 µg/ml), while 0.1 mmol/l (equivalent to 20.84 µg/ml) inhibited the enzyme by 24%. The aldehyde dehydrogenase activity was reduced by 87% at a concentration of 0.1 mmol/l (equivalent to 20.84 µg/ml). All of the values were significantly reduced ($p < 0.05$; Freundt and Schreiner, 1988).

The activity of cytoplasmic aldehyde dehydrogenase from sheep liver was inhibited by up to 60% within 15 min by 0.5 mM tetramethylthiuram monosulfide (equivalent to 104 µg/ml) in vitro at pH levels of 7.3 and 9.3. Mitochondrial aldehyde dehydrogenase activity was inhibited by 12% at pH 9.3 (Kitson, 1976).

A single oral dose of tetramethylthiuram monosulfide (equivalent to 1/5 of the LD_{50}; no further details) led to a significant ($p \leq 0.05$) reduction in cytochrome oxidase activity in the liver, kidneys, heart, cerebrum and spleen of rats as well as a 32% reduction in the activity of succinate dehydrogenase in the heart muscle. The activity of catalase was unaffected (Korablev and Simorot, 1965).

The effect of an intraperitoneal injection of tetramethylthiuram monosulfide on the level of SH-groups in various organs was studied in rats. The animals were given a single dose equivalent to 1/5 of the LD_{50} (LD_{50} value not given), and the SH-content in the blood, heart, brain, liver, kidneys and spleen was determined 60 min after administration. Tetramethylthiuram monosulfide did not affect the content of SH-groups in the various organs (no further details; Korablev and Kurbat, 1965).

The EC_{50} value for the inhibition of cytochrome P-450 activity in rat liver microsomal fraction in vitro was 1.77 µmol/l (equivalent to 0.37 µg/ml; Freundt et al., 1981a, b).

The intraperitoneal administration of 125 mg tetramethylthiuram monosulfide/kg body weight inhibited the activity of dopamine-b-hydroxylase in the hearts of female Charles River rats and Syrian golden hamsters almost completely (ca. 98%) one hour after administration. After 6 h the activity was ca. 20% in the rats and ca. 41% in the hamsters, while after 24 h it was 23 and ca. 100%, respectively, relative to the control animals. In the adrenal glands, the enzyme was inhibited by 100% in the rats and 93% in the hamsters one hour after administration. After 24 h, the enzyme activity was 41% of the control value in the rat and 100% in the hamster. The dopamine and noradrenaline levels, however, remained unaffected (Lippmann and Lloyd, 1969, 1971).

The administration of a single oral dose of 0.12, 2.08 or 4.16 mmol tetramethylthiuram monosulfide/kg body weight (equivalent to 25, 433 and 866 mg/kg body weight) by gavage to groups of 4 female SPF Wistar rats led to a significant ($p \leq 0.05$) increase in epoxide hydrolase activity in the liver at the two higher doses 18 h after administration. The cytosolic and microsomal glutathione-S-transferase activities were slightly, although not significantly, increased (Schreiner and Freundt, 1985).

The administration of tetramethylthiuram monosulfide to female Wistar rats by gavage at doses of from 30 to 1000 mg/kg body weight caused a significant increase in the activity of epoxide hydratase at the high doses (Karakasevic-Stanisic et al., 1981).

The oral administration of 26 or 867 mg tetramethylthiuram monosulfide/kg body weight to female Wistar rats led to a significant ($p < 0.05$) dose-dependent reduction in the incorporation of radiolabelled palmitic acid into the following phospholipids of the endoplasmic reticulum of the liver, 18 h after administration: lysophosphatidyl choline, phosphatidyl choline, sphingomyelin, phosphatidyl inositol, phosphatidyl serine and phosphatidyl ethanolamine. The 867 mg/kg body weight dose inhibited the incorporation almost completely. Each dose and control group consisted of 3 animals (Torres Alanis et al., 1982).

The EC_{50} values for the inhibition of lipid peroxidation (the parameter measured was the synthesis of malondialdehyde) were 1.38 µmol/l (equivalent to 0.28 µg/ml) and 6.08 µmol/l (equivalent to 1.27 µg/ml) for the microsomal and cytosolic fractions of rat liver, respectively (Freundt et al., 1981 b).

Xanthine oxidase activity from rat liver was inhibited by 700 µg/ml (no further details; Matsuoka, 1960).

At a dose range of 20 to 80 mg/kg body weight administered intraperitoneally to mice, tetramethylthiuram monosulfide increased the sleeping time induced by hexobarbital (75 mg/kg body weight by intraperitoneal injection) by a maximum of 2.5-fold, and that induced by ethanol by a maximum of 7-fold (Poitou et al., 1978).

A single intraperitoneal injection of 75 mg hexobarbital/kg body weight led to average sleeping times of ca. 95 and 245 min in the rat 18 h after oral administration of 26 and 867 mg tetramethylthiuram monosulfide/kg body weight, respectively (hexobarbital control 65 min). Paralysis induced by the intraperitoneal administration of 43 mg zoxazolamine/kg body weight (ca. 100 min) was significantly ($p \leq 0.05$) increased to ca. 245 min, 18 h after oral administration of 867 mg tetramethylthiuram monosulfide/kg body weight. The parameter measured was the righting reflex (the ability to assume a standing position after being placed on the back; Torres Alanis et al., 1982).

Tetramethylthiuram monosulfide was not phototoxic in mice. There were no indications of photodermatitis following daily topical application of 1% tetramethylthiuram monosulfide in vaseline to the skin of naked mice (no further details) either with simultaneous UV irradiation (168 h; wavelength 290 to 320 nm) or followed by irradiation for 1 h with a mercury vapour lamp (Muller and Mitchell, 1971).

In vitro, tetramethylthiuram monosulfide alkylates neither NBP (4-(p-nitrobenzyl)-pyridine) nor deoxyguanosine. The concentrations of tetramethylthiuram monosulfide tested were 0.286 and 0.1 µmol/l (equivalent to 59 ng/ml and 20.8 mg/ml, respectively; Hemminki et al., 1980).

8. Experience in humans

Tetramethylthiuram monosulfide can cause skin sensitization in man. The cause of sensitization has often been rubber gloves or preparations containing tetramethylthiuram monosulfide (Kanerva et al., 1987; Lammintausta and Kalimo, 1985; Themido and Brandao, 1984; van Ketel, 1976, 1984).

Cross-reactions to related thiurams may also occur (Baer et al., 1973; Themido and Brandao, 1984).

Patch tests with tetramethylthiuram monosulfide were carried out in accordance with the ICDRG (International Contact Dermatitis Research Group) in 255 patients with dermatitis in Finland between 1980 and 1982. This involved the application of tetramethylthiuram monosulfide under occlusive cover for 24 h (concentration unspecified). Tetramethylthiuram monosulfide led to a positive reaction in 35 of the 255 patients tested after 48, 72 and 96 h (Lammintausta and Kalimo, 1985).

Between 1974 and 1983, 542 patients with contact dermatitis were studied in Finland. In 63 patients, contact dermatitis had been caused by rubber gloves. Patch tests with the ICDRG standard series for rubber chemicals (24-hour occlusive cover) gave positive reactions to tetramethylthiuram monosulfide in 30 of the 63 patients (Estlander et al., 1986; Kanerva et al., 1987).

In Portugal, 128 patients who showed allergic reactions to a thiuram mixture (tetramethylthiuram monosulfide, tetramethylthiuram disulfide, tetraethylthiuram disulfide and dipentamethylthiuram disulfide) were tested for contact allergy to tetramethylthiuram monosulfide between 1977 and 1982. The patch test was positive in 88 subjects and cross-reactions to the above-mentioned thiurams also occurred (no details on concentrations tested; Themido and Brandao, 1984).

Epicutaneous tests conducted in accordance with the ECDRG (European Contact Dermatitis Research Group) at the University Skin Clinic in Heidelberg from 1969 to 1984, in 21 patients with marked allergy to rubber gloves, showed a positive reaction to tetramethylthiuram monosulfide (1% in white vaseline; 24-hour exposure) in one of the patients (4.7%). Of 10 further patients, 2 (20%) reacted positively to tetramethylthiuram monosulfide (Frosch et al., 1987).

Tetramethylthiuram monosulfide (2% in paraffin) caused positive reactions in patch tests in 5 of 200 contact dermatitis patients (Rudzki et al., 1976).

Of 3851 patients patch-tested in accordance with ECDRG at the University of Erlangen-Nürnberg between 1985 and 1990, 145 (3.8%) showed positive reactions to a mixture of rubber chemicals. A series of individual rubber chemicals was tested in 35 of the 145 patients. Of the 35 patients, 15 reacted to tetramethylthiuram monosulfide (0.25% in yellow vaseline; von Hintzenstern et al., 1991).

Positive reactions were seen in 20 patients (10 men and 10 women) from a group of 401 dermatitis patients tested with 1% tetramethylthiuram monosulfide (in vaseline). The cause was assumed to be rubber shoes in 6 of the patients, and a germicidal soap containing tetramethylthiuram monosulfide in 4 others. A patch test with a 1% tetramethylthiuram monosulfide-containing soap solution caused a positive reaction in these 4 patients (Gaul, 1957).

Of 100 contact dermatitis patients patch-tested in Spain, 37 gave positive reactions to tetramethylthiuram monosulfide (2% in vaseline; Conde-Salazar Gomez and Gomez Urcuyo, 1976 a, b).

In 2 of 13 contact dermatitis patients, 1% tetramethylthiuram monosulfide led to a positive reaction in patch tests. The results were evaluated 72 h after application (Kantoh et al., 1985).

Positive reactions to 1% tetramethylthiuram monosulfide applied in anhydrous eucerin were seen in 4 of 5 patients with eczema due to rubber (Bandmann, 1956).

Of 21 patients with dermatosis of the feet due to rubber shoe components, 3 showed a positive reaction to tetramethylthiuram monosulfide (1% in petrolatum; Blank and Miller, 1952).

A 35-year-old woman who had contact dermatitis of the feet due to wearing rubber boots reacted positively to 1 and 0.1% tetramethylthiuram monosulfide in patch tests (English abstract of a Japanese study; Fukuno et al., 1985).

Positive results in ICDRG patch tests in a 50-year-old cleaning woman with allergic reactions to rubber gloves and boots included tetramethylthiuram monosulfide (no further details; Goitre et al., 1981).

Patch tests in a 39-year-old woman with erythema, oedema and peeling of the upper and lower eyelids gave a strong positive reaction to the rubber parts of an eyelash curler and to thiuram mix. Individual tests with rubber chemicals revealed a strong allergic reaction to tetramethylthiuram monosulfide among other chemicals (Cabrita et al., 1991).

The lymphocytes of 4 patients with tetramethylthiuram monosulfide-positive patch tests proliferated after incubation with dimethylthiocarbamoyl-HSA (human serum albumin) conjugates in the lymphocyte transformation test (LTT). The dimethylthiocarbamoyl group was considered to be the relevant antigenic determinant of tetramethylthiuram monosulfide (see Sect. 7.9 for more detailed description of study; Kimber et al., 1991).

Toxic effects in people exposed to tetramethylthiuram monosulfide dust have so far not been reported (no further details; Schunk, 1984).

On inhalation, dermal or oral absorption of tetramethylthiuram monosulfide with subsequent alcohol consumption, it is possible that an ethanol intolerance reaction (antabuse effect) may occur, as with prior tetramethylthiuram disulfide exposure. The reaction is characterized by restlessness, sleep disturbances, tremors of the fingers and tongue, alternating sensations of heat and coldness, nausea, vomiting and stomach pains (Reinl, 1966).

9. Threshold limit values

No information available.

References

Akabane, J., Ikomi, F.
Effects of disulfiram (tetraethylthiuramdisulfide) and its related compounds on the metabolism of alcohol
Med. J. Shinshu Univ., 3, 345–351 (1958)

Baer, R.L., Ramsey, D.L., Biondi, E.
The most common contact allergens
Arch. Dermatol., 108, 74–78 (1973)

Bandmann, H.J.
Vulkanisationsbeschleuniger als Kontaktekzemallergene
Hautarzt, 7 (9), 419–421 (1956)

Bayer AG, Institut für Toxikologie
Vulkacit Thiuram MS, Untersuchungen zur akuten oralen Toxizität an männlichen Wistar-Ratten
Unpublished report (1980)

Bionetics Research Laboratories, Inc., Bethesda, Maryland
Evaluation of carcinogenic, teratogenic, and mutagenic activities of selected pesticides and industrial chemicals. Volume I: Carcinogenic study
Report on contract no. PH43-64–57 and PH43-67–735 (1968 a)
On behalf of the National Cancer Institute
NTIS/PB-223 159

Bionetics Research Laboratories, Inc., Bethesda, Maryland
Evaluation of carcinogenic, teratogenic, and mutagenic activities of selected pesticides and industrial chemicals. Volume II: Teratogenic study in mice and rats
Report on contract no. PH43-64-57 and PH43-67-735 (1968 b)
On behalf of the National Cancer Institute
NTIS/PB-223 160

Blank, I.H., Miller, O.G.
A study of rubber adhesives in shoes as the cause of dermatitis of the feet
JAMA, 149, 1371-1374 (1952)

Cabrita, J.C., Goncalo, M., Azenha, A., Goncalo, S.
Allergic contact dermatitis of the eyelids from rubber chemicals
Contact Dermatitis, 24, 145-146 (1991)

Conde-Salazar Gomez, L., Gomez Urcuyo, J.F.
Sensibilidad a los componentes de la goma en obreros de la construcción (revisión de 100 cabos)
Actas Derma-Sifiliograficas, 67, 297-310 (1976 a)

Conde-Salazar Gomez, L., Gomez Urcuyo, J.F.
Sensibilidad a los componentes de la goma en obreros de la construccón (revisión de 100 cabos)
Medicina Y Seguridad Del Trabajo, 24, 47-54 (1976 b)

Davydov, B.Y., Ratnikova, T.V., Ginak, A.I., Uryupov, O.Y.
Toxicological evaluation of thiopyrophosphates as accelerators of a polyfunctional effect
Kauch. Rezina, issue 9, 53-54 (1980)
Also: Chem. Abstr., 94, 025792p (1981)

Dieke, S.H., Allen, G.S., Richter, C.P.
The acute toxicity of thioureas and related compounds to wild and domestic Norway rats
J. Pharmacol. Exp. Ther., 90, 260-270 (1947)

Donner, M.
Use of Drosophila as indicator of risk chemicals in rubber factory
SGF Publ., 57, 47–53 (1981)

Donner, M., Husgafvel-Pursiainen, K., Jenssen, D., Rannug, A.
Mutagenicity of rubber additives and curing fumes
Scand. J. Work Environ. Health, 9, Suppl. 2, 27–37 (1983)

Doull, J., Plzak, V., Brois, S.J.
A survey of compounds for radiation protection
Armed Services Technical Information Agency, Arlington, VA
NTIS Report AD 277–689 (1962)

Estlander, T., Jolanki, R., Kanerva, L.
Dermatitis and urticaria from rubber and plastic gloves
Contact Dermatitis, 14, 20–25 (1986)

FhG (Fraunhofer-Institut für Toxikologie und Aerosolforschung, Schmallenberg)
Bericht über die Prüfung von Vulkacit Thiuram MS auf primäre Hautreizwirkung
Unpublished report (1980 a)
On behalf of Bayer AG

FhG (Fraunhofer-Institut für Toxikologie und Aerosolforschung, Schmallenberg)
Bericht über die Prüfung von Vulkacit Thiuram MS auf Schleimhautreizwirkung
Unpublished report (1980 b)
On behalf of Bayer AG

Freundt, K.J., Leyck, S., Kamal, A.A.M., Römer, K.G.
Influence of CS-containing compounds on hepatic lipids in experimental animals
G. Ital. Med. Lav., 3, 117–119 (1981 a)

Freundt, K.J., Römer, K.G., Kamal, A.A.M.
The inhibitory action of dithiocarbamates and carbon disulphide on

malondialdehye formation resulting from lipid peroxidation in rat liver microsomes
J. Appl. Toxicol., 1, 215–219 (1981 b)

Freundt, K.J., Schreiner, E.
In vitro inactivation of hepatic alcohol dehydrogenase and aldehyde dehydrogenases from rats by dithiocarbamates with or without metals
Bull. Environ. Contam. Toxicol., 41, 540–547 (1988)

Frosch, P.J., Born, C.M., Schütz, R.
Kontaktallergien auf Gummi-, Operations- und Vinylhandschuhe
Hautarzt, 38, 210–217 (1987)

Fukuno, M., Tateishi, H., Haneda, T., Sagami, S.
A case of rubber dermatitis
Skin Res., 27, 787–792 (1985)

Garcia de Torres, G., Römer, K.G., Torres Alanis, O., Freundt, K.J.
Blood acetaldehyde levels in alcohol-dosed rats after treatment with ANIT, ANTU, dithiocarbamate derivatives, or cyanamide
Drug Chem. Toxicol., 6, 317–328 (1983)

Gaul, L.E.
Results of patch testing with rubber antioxidants and accelerators
J. Invest. Dermatol., 29, 105–110 (1957)

Goitre, M., Bedello, P.G., Cane, D.
Allergic dermatitis and oral challenge to tetramethylthiuram disulphide
Contact Dermatitis, 7, 272–273 (1981)

Goodyear, Health, Safety and Environmental Section
Mutagenicity evaluation of Unads – Ames test results
Unpublished report, laboratory report no. 79–62 (1980)

Goodyear, Fiber & Polymer Products Research Div.
The mutagenic evaluation of Unads with *Drosophila*
Unpublished report (1982)

Hald, J., Jacobsen, E., Larsen, V.
The antabuse effect of some compound. Related to antabuse and cyanamide
Acta Pharmacol. Toxicol., 8, 329–337 (1952)

Hedenstedt, A.
Mutagenicity of rubber – vulcanization gases
SGF Publ., 57, 39–46 (1981)

Hedenstedt, A., Rannug, U., Ramel, C., Wachtmeister, C.A.
Mutagenicity and metabolism studies on 12 thiuram and dithiocarbamate compounds used as accelerators in the Swedish rubber industry
Mutat. Res., 68, 313–325 (1979)

Heinrich-Hirsch, B.
Possible contribution of in vitro methods assessment; discussion of the presentation
In: Neubert et al., p. 471 (1992)

Hemminki, K., Falck, K., Vainio, H.
Comparison of alkylation rates and mutagenicity of directly acting industrial and laboratory chemicals
Arch. Toxicol., 46, 277–285 (1980)

Hintzenstern, von, J., Heese, A., Koch, H.U., Peters, K.P., Hornstein, O.P.
Frequency, spectrum and occupational relevance of type IV allergies to rubber chemicals
Contact Dermatitis, 24, 244–252 (1991)

Kanerva, L., Estlander, T., Jolanki, R.
Berufsbedingte Dermatosen, verursacht durch Gummi- und Plastikhandschuhe (German translation of the Russian)
Gig. Tr. Prof. Zabol., issue 4, 11–13 (1987)

Kantoh, H., Ishihara, M., Itoh, M., Hosono, K., Nishimura, M.
Allergens in rubber products
Hifu, 27, 501–509 (1985)

Karakasevic-Stanisic, B., Schreiner, E., Freundt, K.J.
Enhanced epoxide hydratase activity by antioxidants of the dithiocarbamate type
J. Cancer Res. Clin. Oncol., 99, A32-A33 (1981)

Ketel, van, W.G.
Thiuram-mix
Contact Dermatitis, 2, 232–233 (1976)

Ketal, van, W.G.
Contact urticaria from rubber gloves after dermatitis from thiurams
Contact Dermatitis, 11, 323–324 (1984)

Kimber, I., Quirke, S., Cumberbatch, M., Ashby, J., Paton, D., Aldridge, R.D., Hunter, J.A.A., Beck, M.H.
Lymphocyte transformation and thiuram sensitization
Contact Dermatitis, 24, 164–171 (1991)

Kitson, T.M.
The effect of some analogues of disulfiram on the aldehyde dehydrogenases of sheep liver
Biochem. J., 155, 445–448 (1976)

Korablev, M.V.
The toxicity of dithiocarbamic acid derivatives and of compounds structurally similar to them
Farmakol. Toksikol., 28, 230–233 (1965)

Korablev, M.V.
The toxicological characteristics of dithiocarbamic acid derivatives used in the national economy and in the medicine; a review
Farmakol. Toksikol., 32, 256–362 (1969)

Korablev, M.V., Evets, M.A.
Antidiuretic action of the dithiocarbaminic acid derivatives
Farmakol. Toksikol., 40, 603–606 (1977)

Korablev, M.V., Kurbat, N.M.
Effect of derivatives of dithiocarbamic acid on the level of sulfhydryl groups in organs and tissues of rats
Zdravookhr. Belorussii, 3, 34–35 (1965)
Also: Chem. Abstr., 64, 2635a (1966)

Korablev, M.V., Simorot, R.P.
The influence exerted by dithiocarbaminic acid derivatives on the enzymatic activity of cytochromoxidase, succino-dehydrogenase and catalase
Farmakol. Toksikol., 28, 234–237 (1965)
Also: Chem. Abstr., 63, 17013 h (1965)

Korhonen, A., Hemminki, K., Vainio, H.
Application of the chicken embryo in testing for embryotoxicity – thiurams
Scand. J. Work Environ. Health, 8, 63–69 (1982)

Korhonen, A., Hemminki, K., Vainio, H.
Toxicity of rubber chemicals towards three-day chicken embryos
Scand. J. Work Environ. Health, 9, 115–119 (1983)

Lammintausta, K., Kalimo, K.
Sensitivity to rubber – study with rubber mixes and individual rubber chemicals
Derm. Beruf Umwelt, 33, 204–208 (1985)

LBI (Litton Bionetics, Inc.)
Mutagenicity evaluation of BIO-76–275 monothiurad
Unpublished report (1976)
On behalf of Monsanto Company, St. Louis, Missouri

LBI (Litton Bionetics, Inc.)
Mutagenicity evaluation of BIO-77–323 monothiurad in the mouse lymphoma forward mutation assay
Unpublished report (1978)
On behalf of Monsanto Company, St. Louis, Missouri

LBI (Litton Bionetics, Inc.)
Evaluation of TMTM in the in vitro transformation of BALB/3T3 cells assay
Unpublished report, LBI project no. 20992 (1981)
On behalf of Goodyear Tire and Rubber, Co., Akron, Ohio

Lippmann, W., Lloyd, K.
Diethyldithiocarbamate and structurally-related compounds and the uptake and release of noradrenaline in the heart of the rat
J. Pharm. Pharmacol., 21, 780–781 (1969)

Lippmann, W., Lloyd, K.
Effects of tetramethylthiuram disulfide and structurally-related compounds on the dopamine-b-hydroxylase activity in the rat and hamster
Arch. Int. Pharmacodyn. Ther., 189, 348–357 (1971)

Matsuoka, Y.
Effects of disulfiram and its related compounds on the activity of xanthine oxidase and catalase
Shinshu Igaku Zasshi, 9, 497–506 (1960)
Also: Chem. Abstr., 56, 14873a (1962)

McCormick, W.E.
Environmental health control for the rubber industry
Rubber Chem., 44, 512–533 (1971)

Mitani, C.
Pharmacological studies on thiurams, dithiocarbamates, guanidine derivatives, and p-quinone dioxime, especially effects of these compounds on alcohol metabolism. I. Effects on the metabolism of alcohol
Shinshu Igaku Zasshi, 9, 90–101 (1960)
Also: Chem. Abstr., 56, 14887e (1962)

Monsanto Company
Material safety data – Mono Thiurad accelerator (undated)

Muller, R., Mitchell, J.C.
Psoralen-type phototoxicity of tetramethyl-thiuram-monosulphide for *Candida albicans*; not for mouse or man
J. Invest. Dermatol., 56, 340 (1971)

Neubert, D., Kavlock, R.J., Merker, H.J., Klein, J. (eds.)
Risk assessment of prenatally-induced adverse health effects
Springer-Verlag (1992)

Neubert, D.
Personal communication (1993)

Poitou, P., Marignac, B., Gradiski, D.
Effet sur la système nerveux central du monosulfure et du disulfure de tétraméthylthiurame, comparison avec le disulfirame
J. Pharmacol., 9, 35–44 (1978)

Rannug, A., Rannug, U., Ramel, C.
Genotoxic effects of additives in synthetic elastomers with special consideration to the mechanism of action of thiurams and dithiocarbamates
Prog. Clin. Biol. Res., 141, issue: Ind. Hazards Plast. Synth. Elastomers, 407–419 (1984)

Reinl, W.
Alkoholüberempfindlichkeit nach Umgang mit dem Fungicid Tetramethylthiuram-disulfid (TMTD)
Arch. Toxicol., 22, 12–15 (1966)

Römer, K.G., Freundt, K.J., Garcia de Torres, G., Torres Alanis, O.
Kombinationswirkungen: Thiurame und Alkohol – tierexperimenteller Zugang, biometrische Analyse und Interpretation
Komb. Belastungen Arbeitsplatz, Ber. Jahrestag Dtsch. Ges. Arbeitsmed., 22, 205–209 (1982)

Römer, K.G., Torres Alanis, O., Garcia de Torres, G., Freundt, K.J.
Delayed ethanol elimination from rat blood after treatment with thi-

ram, tetramethylthiuram monosulfide, ziram, or cyanamide
Bull. Environ. Contam. Toxicol., 32, 537–542 (1984)

Rudzki, E., Ostaszewski, K., Grzywa, Z., Kozlowska, A.
Sensitivity to some rubber additives
Contact Dermatitis, 2, 24–27 (1976)

Schreiner, E., Freundt, K.J.
Activities of hepatic epoxide hydrolase and glutathione S-transferase in rats under the influence of tetramethyl thiuramdisulfide, tetramethyl thiurammonosulfide or dimethyl dithiocarbamate
Toxicol. Lett., 25, 147–152 (1985)

Schubart, R.
Dithiocarbamic acid and derivatives
In: Ullmann's encyclopedia of industrial chemistry
5th ed., vol. A9, p. 6–7
VCH Verlagsgesellschaft, Weinheim (1987)

Schunk, W.
Die Gefährdung von Arbeitern in der Gummiindustrie durch chemische Einsatzstoffe
Gummi, Fasern, Kunststoffe, 37, 58–64 (1984)

Skofitsch, G.
Vom Hund zum Ei? Grobe und kleine Laboratoriumstiere
In: Lembeck, F. (ed.)
Alternativen zum Tierversuch, p. 71–79
Georg Thieme Verlag, Stuttgart, New York (1988)

Takahashi, A.
Problems of hygiene maintenance for food coming into contact with rubber and plastics products (English translation of the Japanese)
Nippon Gomu Kyokaishi, 48 (9), 573 (1975)
Translation in: Int. Polymer Sci. Tech., 3 (1), T/93-T/105 (1976)

Themido, R., Brandao, F.M.
Contact allergy to thiurams
Contact Dermatitis, 10, 251 (1984)

Torres Alanis, O., Freundt, K.J., Liebaldt, G.P.
Toxicity studies on tetramethylthiuram monosulfide
Environ. Res., 28, 199–211 (1982)

Uniroyal
Bioassay: Ames Salmonella microsome
Unpublished report (1984)

Vorob'eva, R.S.
Comparative toxicity of tetramethylthiuram monosulfide and tetramethylthiuram disulfide
Kauch. Rezina, 23, 34–36 (1964)
Also: Chem. Abstr., 61, 7586f (1964)

Vorob'eva, R.S.
Comparative toxicity of thiuram compounds, depending on the number of sulfide groups
Gig. Tr. Prof. Zabol., 9, 53–56 (1965)
Also: Chem. Abstr., 64, 10292f (1966)

You, X., Zhou, Y., Hu, Y.
Mutagenicity of fourteen rubber accelerators
Huanjing Kexue, 3, 39–42 (1982)

Di-2-ethylhexylamine

1. Summary and assessment

Di-2-ethylhexylamine is moderately toxic after acute oral administration to rats or a single dermal application to rabbits (LD_{50} rat oral ca. 1000 and 1640 mg/kg body weight; LD_{50} rabbit dermal 958 mg/kg body weight). After intraperitoneal injection in the mouse, di-2-ethylhexylamine is clearly more toxic (LD_{50} ca. 50 mg/kg body weight). In this case, the severe local irritant effect could play a role, and it is presumably also responsible for the higher toxicity on inhalation (4-hour LC_{50} 0.91 mg/l; substance tested as an aerosol).

The substance is severely irritating or corrosive to rabbit skin and causes serious damage to the rabbit eye.

Di-2-ethylhexylamine is not mutagenic in the Salmonella/microsome test in strains TA 100, TA 1535, TA 97 or TA 98 either with or without metabolic activation.

2. Name of substance

2.1	Usual name	Di-2-ethylhexylamine
2.2	IUPAC-name	Di-(2-ethylhexyl)amine
2.3	CAS-No.	106–20–7
2.4	EINECS-No.	203–372–4

3. Synonyms, common and trade names

Bis(2-ethylhexyl)-amine
2,2′-Diethyl-dihexylamine
Diisooctylamine

2-Ethyl-N-(2-ethylhexyl)-1-hexanamine
Di-2-ethylhexylamin

4. Structural and molecular formulae

4.1 Structural formula $(CH_3\text{-}CH_2\text{-}CH_2\text{-}CH_2\text{-}CH(C_3H_5)\text{-}CH_2)_2N\text{-}H$

$$(CH_3\text{-}CH_2\text{-}CH_2\text{-}CH_2\text{-}\underset{\displaystyle C_3H_5}{\underset{|}{CH}}\text{-}CH_2)_2N\text{-}H$$

4.2 Molecular formula $C_{16}H_{35}N$

5. Physical and chemical properties

5.1	Molecular mass, g/mol	241.46
5.2	Melting point, °C	< –70 (BASF, 1991)
5.3	Boiling point, °C	157–159 (at 27 hPa) (BASF, 1991)
5.4	Vapour pressure, hPa	< 0.01 (at 20 °C) 0.1 (at 55 °C) (BASF, 1991)
5.5	Density, g/cm3	0.806 (at 20 °C) (BASF, 1991)
5.6	Solubility in water	Insoluble (BASF, 1991)
5.7	Solubility in organic solvents	Miscible with almost all common solvents (BASF, 1987)
5.8	Solubility in fat	No information available
5.9	pH value	Suspension reacts as alkali (BASF, 1985)
5.10	Conversion factor	1 ml/m³ (ppm) ≙ 9.86 mg/m³ 1 mg/m³ ≙ 0.10 ml/m³ (ppm) (at 1013 hPa and 25 °C)

6. Uses

Intermediate in the chemical industry; in the manufacture of textile processing aids, dyes, pesticides, polycondensates, corrosion inhibitors and petroleum additives; stabilizer for organic halogen compounds (BASF, 1987).

7. Experimental results

7.1 Toxicokinetics and metabolism

No information available.

7.2 Acute and subacute toxicity

The acute oral toxicity of di-2-ethylhexylamine was studied in tests to obtain "range-finding toxicity data". Groups of 5 rats (sex unspecified) received 4 dosages in a geometrical series, with a ratio of 2 between successive dose levels and were observed for 14 days. The LD_{50} was 1640 (1440 to 1870) mg/kg body weight (Smyth et al., 1949).

In a further acute oral toxicity study, male and female rats (group size unspecified) received di-2-ethylhexylamine as a 20% aqueous emulsion in traganth by gavage. The LD_{50} was ca. 1000 mg/kg body weight after a 14-day observation period. Signs of toxicity included dyspnoea, mild apathy and diarrhoea. Post-mortems revealed atony of the gut (BASF, 1967).

A dermal LD_{50} value of 1.19 ml/kg, equivalent to 958 mg/kg, was reported for rabbits, with no further details provided (UCC, 1968).

To determine the intraperitoneal LD_{50}, male and female mice (group size unspecified) were injected in the peritoneal cavity with di-2-ethylhexylamine as a 1% emulsion in traganth. The observation period was 14 days. The LD_{50} was ca. 48 mg/kg body weight. Signs of toxicity observed included dyspnoea, reeling, tremors and spasms. Post-mortems revealed atony of the gut with adhesions and concrescence in the abdominal cavity (BASF, 1967).

In an inhalation hazard test, 6 male rats were acutely exposed to an atmosphere saturated with di-2-ethylhexylamine at 20 °C for 8 h.

The observation period was 14 days. All of the rats survived. No details of signs of toxicity or of the findings at autopsy were provided (Smyth et al., 1949).

In a further inhalation hazard test at 20 °C, 6 male and 6 female rats similarly survived a single 8-hour exposure and 14-day observation period. Symptoms described included signs of irritation of the mucous membranes and post-mortems revealed occasional bronchopneumonia (BASF, 1967).

The acute inhalation toxicity of 99.6% pure di-2-ethylhexylamine was studied in groups of 5 male and 5 female Wistar rats (average initial weights 260 and 181 g, respectively), in accordance with OECD guideline no. 403. The analytically determined concentrations were 0.28, 0.53, 0.84, 1.28 and 2.85 mg/l air as an aerosol with a mass median aerodynamic diameter of 1.4 to 1.8 mm and a geometric standard deviation of between 3.3 and 3.8. Exposure was for a single 4-hour period, and the observation period was 14 days. The LC_{50} value was determined as 0.91 mg/l for male and female rats. The main symptom observed was local irritation of the airways and of the eyes. Autopsies on the rats that died revealed local hyperaemia of the lungs and moderate pulmonary emphysema. There were no pathological findings in post-mortems carried out at the end of the observation period (BASF, 1991).

7.3 Skin and mucous membrane effects

The skin irritancy of di-2-ethylhexylamine was initially studied in 5 rabbits during tests to obtain "range-finding toxicity data" and an evaluation based on the method of Draize was undertaken. After non-occlusive exposure for 24 h, neat di-2-ethylhexylamine gave an irritation index of 5, that is it led to severe erythema, oedema or slight necrosis (Smyth et al., 1949).

In a further study (no further details), the neat substance led to redness and swelling when applied to the abdominal skin of rabbits (UCC, 1968).

A third study produced reddening and severe, prominent oedema after application of the neat substance to the backs of rabbits (2 animals) for 1, 5 and 15 min. One week later, mild redness, rhagades (cracks in the skin) and mild (1 and 5 min) or severe (15 min) necrosis and induration were evident at the application site. After an exposure period of 20 h, neat di-2-ethylhexylamine led to mild redness, severe oedema and severe necrosis of the dorsal skin of the rabbit. Severe necrosis and induration at the application site were recorded 8 days later (BASF, 1967).

The eye irritancy of di-2-ethylhexylamine was also first studied in rabbits during the acquisition of range-finding toxicity data. The substance led to an "injury grade" of 8, that is 5 ml of a 15% solution caused an irritation value of over 5.0 and 5 µl of a 5% solution caused one of not more than 5.0 (index 5.0 = marked irritation of the cornea and iris; Smyth et al., 1949; Carpenter and Smyth, 1946).

In a further eye irritation study in 2 rabbits, the instillation of 50 µl of the neat product caused severe redness, severe oedema and slight clouding of the cornea (which had still not cleared up after 8 days), after 1 and 24 h (BASF, 1967).

7.4 Sensitization

No information available.

7.5 Subchronic and chronic toxicity

No information available.

7.6 Genotoxicity

7.6.1 In vitro

Di-2-ethylhexylamine (purity unspecified) was tested in the Salmonella/microsome test in strains TA 100, TA 1535, TA 97 and TA 98 with and without metabolic activation. The metabolizing system was S9-mix from Aroclor 1254-induced hamster and rat liver and the concentrations tested were 3 to 6666 µg/plate. Cytotoxicity was observed from 166 µg/plate. Di-2-ethylhexylamine was not mutagenic in this study (Zeiger et al., 1988).

7.6.2 In vivo

No information available.

7.7 Carcinogenicity

No information available.

7.8 Reproductive toxicity

No information available.

7.9 Effects on the immune system

No information available.

7.10 Neurotoxicity

No information available.

7.11 Other effects

In vivo, di-2-ethylhexylamine increases the penetration of drugs dissolved in it through the skin. This effect is not caused by capillary permeability or accelerated transport away from the skin, but solely by increased permeability of the stratum corneum (Creasey et al., 1971).

8. Experience in humans

No information available.

9. Threshold limit values

No information available.

References

BASF AG, Gewerbehygienisch-Pharmakologisches Institut
Di-(2-äthylhexyl)-amin – Ergebnis der gewerbetoxikologischen Vorprüfung
Unpublished report (1967)

BASF AG
DIN-Safety data sheet "Di-2-ethylhexylamine", 10/85 (1985)

BASF AG
Data sheet "Di-2-ethylhexylamin" (1987)

BASF AG
AIDA-Grunddatensatz 1-Hexanamine, 2-ethyl-N-(2-ethylhexyl)-(1991)

BASF AG, Abteilung Toxikologie
Study on the acute inhalation toxicity LC_{50} of di-2-ethylhexylamin as a liquid aerosol in rats, 4 hour exposure
Unpublished report, project no. 13I0722/897066 (1991)
On behalf of BG Chemie

Carpenter, C.P., Smyth, H.F., Jr.
Chemical burns of the rabbit cornea
Am. J. Ophthalmol., 29, 1363–1372 (1946)

Creasey, N.H., Allenby, A.C., Schock, C.
Mechanism of action of accelerants. The effect of the cutaneously applied penetration accelerants on the skin circulation of the rat
Br. J. Dermatol., 85, 368–380 (1971)

Smyth, H.F., jr., Carpenter, C.P., Weil, C.S.
Range-finding toxicity data, list III
J. Ind. Hyg. Toxicol., 31, 60–62 (1949)

UCC (Union Carbide Corporation)
Data sheet "Toxicology studies di-2-ethylhexyl amine" (1968)

Zeiger, E., Anderson, B., Haworth, S., Lawlor, T., Mortelmans, K.
Salmonella mutagenicity tests: IV. Results from the testing of 300 chemicals
Environ. Mol. Mutagen., 11, Suppl. 12, 1–158 (1988).

p-Chloroaniline

1. Summary and assessment

After a single oral dose, p-chloroaniline is rapidly and almost completely absorbed by the organism. The main metabolites in the blood are p-chloroacetanilide and 2-amino-5-chlorophenyl sulfate. The radioactivity reaches a maximum in the plasma one hour after administration of ^{14}C-labelled p-chloroaniline, and is enriched in the erythrocytes within 24 h. In the urine of mice, rats and monkeys, 80, 81 to 87 and 55%, respectively of p-chloroaniline and its metabolites are excreted within 24 h, while in the faeces, only 5.2, 8.3 and 1.2%, respectively are excreted within 4 days. The main metabolite in the urine at 49% (mice), 54% (rats) and 36% (monkeys) is 2-amino-5-chlorophenyl sulfate. The total excretion in 24-hour urine is 80% in mice, 88% in rats and 56% in monkeys, and the total excretion after 4 days (urine and faeces) is 91% (mice), 101% (rats) and 82% (monkeys). p-Chloroaniline is eliminated from the tissues in 2 phases with half-lives of less than 10 minutes and 3 to 4 h, respectively. In humans, p-chloroaniline is mainly excreted in the urine. Conjugated p-chloroaniline and conjugated 2-amino-5-chlorophenol have also been found as major metabolites in the urine in a case of acute poisoning.

Like aniline, p-chloroaniline primarily damages the erythrocytes, causing methaemoglobinaemia with cyanosis and its secondary effects (anaemia, increases in Heinz bodies, reticulocytosis). In this respect it appears to have even more severe effects than aniline. Kidney damage due to p-chloroaniline has also been described in rats.

p-Chloroaniline is of moderate acute toxicity on oral and dermal administration to rats (LD_{50} oral 300 to 425 mg/kg body weight; LD_{50} dermal 335 mg/kg body weight) and on dermal administration to rabbits (LD_{50} dermal 360 mg/kg body weight). Based on its acute dermal toxicity, it appears to be well absorbed through the skin. The acute inhalation toxicity in the rat is 2.34 mg/l (4-hour LC_{50}). On subacute oral administration for 28 days, the tolerable no effect dose for rats is less than 10 mg/kg body weight; the no effect level for increases in Heinz bodies in rats is 1 mg/m^3 air after inhalation exposure for a month.

According to early studies, p-chloroaniline is irritating to the skin and eyes of the rabbit, while other studies which have been carried out in accordance with OECD guidelines suggest that it is not.

p-Chloroaniline is a weak sensitizer in the maximization test in guinea pigs.

After oral administration to rats and mice for 3 months, dose-dependent methaemoglobinaemia occurred down to the lowest doses tested of 7.5 mg/kg body weight/day in the mouse and 5 mg/kg body weight/day in the rat. This was associated with secondary anaemia as well as histopathological effects on the liver, spleen and kidneys which could be attributed to these blood effects. In 6-month subchronic inhalation studies in rats, 15 mg/m^3 was toxic and 1.5 mg/m^3 was the "threshold dose", while on exposure for 3 months, 0.15 mg/m^3 was without effect. The dose having an effect in dogs after oral administration for 3 months was 5 to 15 mg/kg body weight/day (a no effect level was not established). The findings in this study were similar to those described above.

The genotoxicity data are contradictory. p-Chloroaniline is mainly negative in the Salmonella/microsome test up to the cytotoxic concentration of 1000 µg/plate. The same applies to reversion tests in *Escherichia coli*. Other genotoxicity tests (Pol A test in *Escherichia coli*, mitotic recombination in *Saccharomyces cerevisiae* and reversion tests in *Aspergillus nidulans*) have sometimes been positive and sometimes negative. In rat hepatocytes in vitro, p-chloroaniline has shown both positive and negative results, but was negative in the UDS test. p-Chloroaniline is positive in the mouse lymphoma test, but does not cause DNA strand breaks in mouse lymphoma cells. No clear results have been obtained in tests for chromosome aberrations and sister chromatid exchange in Chinese hamster ovary (CHO) cells. Overall, the substance should be evaluated as being genotoxic, based on the in vitro studies. In vivo, p-chloroaniline is negative in the micronucleus test in the mouse at the maximum tolerated oral dose (180 mg/kg body weight) and positive in the somatic mutation and recombination test in *Drosophila melanogaster*.

With regard to the carcinogenicity of p-chloroaniline, several studies are available in rats and mice, with both negative and positive results. A 78-week feeding study in rats and mice gave no clear results, while a 103-week gavage study produced clear evidence of carcinogenic activity in male rats and no evidence of carcinogenic activity in female mice. The evidence in female rats was equivocal, while there were some indications of carcinogenic activity in male mice. All of the studies primarily involved partly-metastasized spleen tumours (fibromas, fibrosarcomas, sarcomas, carcinomas, haemangiosarcomas), as well as liver tumours in the mice (hepatocellular adenomas, carcinomas, haemangiosarcomas). These tumours, the spontaneous incidence rate of which is very low, are also seen after chronic administration of aniline, and are viewed as a consequence of the haematotoxic effect of p-chloroaniline, although a genotoxic mechanism for tumour induction cannot be excluded.

p-Chloroaniline was mainly positive in four cell transformation tests.

In vitro studies with human skin (from surgical procedures) show that p-chloroaniline is well absorbed by intact skin, and even more so by abraded skin.

In humans as well as animals, the acute effect of p-chloroaniline (methaemoglobin formation) is more severe than that of aniline. Alcohol can lead to augmentation of the effect.

With regard to the acute toxicity of p-chloroaniline, case reports are available on industrial poisonings, one of which was fatal. One patient survived after appropriate therapy (vitamin C, methylene blue), despite a maximum methaemoglobin level of 70%.

It is reported from a production unit that p-chloroaniline concentrations of ca. 60 mg/m^3 air have led to methaemoglobin concentrations of 1.6 to 2.8 g/100 ml blood, to cyanosis, anaemia and even to acute intoxication. However, in this case, dermal absorption of the substance was thought to have contributed to the effects. Other reports indicate that a p-chloroaniline concentration of 44 mg/m^3 air leads to severe toxic effects in man after short exposures, and 22 mg/m^3 air leads to signs of illness on longer exposure.

p-Chloroaniline has a strong tendency to form haemoglobin adducts, and their determination can therefore be used in biomonitoring of employees exposed to the effects of p-chloroaniline in the workplace.

In the European Union, p-chloroaniline is classified in category C2 (substances that should be regarded as carcinogenic in humans).

2. Name of substance

2.1	Usual name	p-Chloroaniline
2.2	IUPAC-name	4-Chloroaniline
2.3	CAS-No.	106–47–8
2.4	EINECS-No.	203–401–0

3. Synonyms, common and trade names

1-Amino-4-chlorobenzene

p-Chloranilin

1-Chloro-4-aminobenzene

4-Chloroaminobenzene

4-Chlorophenylamine

4-Chlorobenzenamine

PCA

4. Structural and molecular formulae

4.1 Structural formula

4.2	Molecular formula	C_6H_6NCl

5. Physical and chemical properties

5.1	Molecular mass, g/mol	127.57
5.2	Melting point, °C	72.5 (Lawrence and Marshall, 1985)
5.3	Boiling point, °C	232 (at 1013 hPa) (Lawrence and Marshall, 1985)
5.4	Vapour pressure, hPa	0.02 (at 20 °C) 0.06 (at 30 °C) 1.33 (at 59 °C) (Hoechst, undated)
5.5	Density, g/cm^3	1.32 (at 20 °C) (Lawrence and Marshall, 1985)
5.6	Solubility in water	Dissolves with difficulty in hot water (Hoechst, undated)
5.7	Solubility in organic solvents	Dissolves readily in alcohol, ether, acetone, benzene, dioxane, chloroform (Hoechst, undated; Lawrence and Marshall, 1985)
5.8	Solubility in fat	No information available
5.9	pH value	No information available
5.10	Conversion factor	1 ml/m3 (ppm) ≙ 5.29 mg/m^3 1 mg/m^3 ≙ 0.19 ml/m^3 (ppm) (at 1013 hPa and 25 °C)

6. Uses

Intermediate in the manufacture of pesticides, isocyanates, dyestuffs and pharmaceuticals (Lawrence and Marshall, 1985).

7. Experimental results

7.1 Toxicokinetics and metabolism

2 male rhesus monkeys (4.0 ± 0.3 kg) were given a single dose of 20 mg > 98% pure [^{14}C]-p-chloroaniline/kg body weight as the hydrochloride intragastrally by nasal intubation. The radioactivity in whole venous blood and in the plasma was determined before administration and 5, 10, 30 min and 1, 2, 4, 8, 12 and 24 h after dosing. The maximum radioactivity in the plasma was reached after 1 h (ca. 160 µM-equivalents), indicating rapid absorption. Within 24 h, the radioactivity was enriched in the erythrocytes. p-Chloroaniline itself similarly reached its highest peak after about an hour, but constituted only about 50% of the total activity at this time and was soon no longer detectable (half-life ca. 1 h). The main metabolite in the circulating blood after one hour was p-chloroacetanilide at ca. 26% of the activity, reaching more than 90% of the activity after 24 h. After 1 h, 2-amino-5-chlorophenyl sulfate (27%) could also be detected. At the time of the highest plasma activity (after 1 h), p-chloroaniline, p-chloroacetanilide and 2-amino-5-chlorophenyl sulfate together constituted 59% of the circulating activity. p-Chloroaniline is therefore rapidly metabolized, with p-chloroacetanilide being the predominant plasma metabolite (Ehlhardt and Howbert, 1991).

2 female C3H-mice (20 ± 1 g), 3 male F344-rats (250 ± 20 g) and 2 male rhesus monkeys (4.0 ± 0.3 kg) were given a single dose of 20 mg > 98% pure [^{14}C]-p-chloroaniline/kg body weight as the hydrochloride intragastrally by gavage (mice and rats) or by nasal intubation (monkeys). The urine and faeces were then analyzed separately for radioactivity. The mice excreted 80% of the radioactivity in the urine within 24 h and 83.4% within 72 h, while the rats excreted 87 and 89.9%, respectively within the same periods. Excretion in the urine in

the monkeys was on average 55% within 24 h, 71% after 2 days and 81% after 4 days. The excretion in the faeces was markedly lower and reached on average only 5.2% in the mice, 8.3% in the rats and 1.2% in the monkeys after 4 days. The recovery rates in the urine (including cage washings) and in the faeces totalled 91% in the mice, 101% in the rats and 82% in the monkeys. The main metabolite in the urine was 2-amino-5-chlorophenyl sulfate at 49% of the activity in mice, 54% in rats and 36% in monkeys, followed by p-chloro-oxanilic acid (oxalic acid-p-chloroanilide) at 6.6% (mice), 11% (rats) and 1% (monkeys). Unidentified compounds accounted for 22% (mice) and 14% (rats and monkeys). Unchanged p-chloroaniline was excreted in the urine at only 1.7% (mice), 0.2% (rats) and 2.5% (monkeys). The total excretion of metabolites, unchanged p-chloroaniline and unidentified compounds in the 24-hour urine was 80% in mice, 88% in rats and 56% in monkeys (Ehlhardt and Howbert, 1991).

Fischer-344 rats were given a single dose of 0.3, 3 or 30 mg ^{14}C-labelled p-chloroaniline/kg body weight by gavage. After 24 h, 81% of the radioactivity had been excreted in the urine and 10% in the faeces. After 7 days, 87% had been eliminated in the urine and 11% in the faeces. After one week, 0.9 to 2.3% of the administered radioactivity was found in the erythrocytes depending on the dose. The remaining tissues contained less than 1%. Accumulation of radioactivity in the erythrocytes was also seen in rats after a single intravenous injection of 3 mg ^{14}C-p-chloroaniline/kg body weight. The erythrocyte/plasma ratio was 2:1 after 2 h, 20:1 after 12 h and 74:1 after 2 days. Elimination from the tissues was biphasic, the first stage having a half-life of less than 10 min, the second stage, one of 3 to 4 h. Similar results were found in studies in dogs involving single intravenous injections of 3 mg/kg body weight (Perry et al., 1981; NTP, 1989).

The p-chloroaniline level in the blood of dogs after a single intravenous injection of 25 mg/kg body weight was just under 18 mg/ml about 5 min after the injection, ca. 13 µg/ml after about 10 min, almost 9 µg/ml after 30 min, ca. 6.5 µg/ml after 1 h, 4 µg/ml after 2 h and ca. 2.5 µg/ml after 3 h (Kiese, 1963).

In studies in isolated bovine lens capsules and rabbit eyes, p-chloroaniline penetrated into all tissues of the eye. After application to the surface of the lens, 17% of the applied dose had penetrated after 3 h and 44% after 24 h (Ismail et al., 1976, 1977).

2-Amino-5-chlorophenol was detected as a metabolite in the 24-hour urine of rabbits (2 to 3 kg) by paper chromatography following a single oral dose of 100 mg/kg body weight (Bray et al., 1956).

Even in earlier studies, 2-amino-5-chlorophenol had been found in the urine of rabbits following acid hydrolysis after oral administration of p-chloroaniline (dose unspecified; Elson et al., 1948).

N-Hydroxyl derivatives of p-chloroaniline were found in the urine of rabbits after a single intravenous dose of 78 mg/kg body weight, 0.4% as hydroxylamine. In dogs, the proportion of hydroxylamine in the urine was 0.04% of the hydroxyl-derivatives, after a single intravenous injection of 39 mg/kg body weight. No hydroxyl-derivatives were detected in the bile in either species (von Jagow et al., 1966).

In other metabolism studies in rabbits given a single intraperitoneal injection of 50 mg/kg body weight, the urine contained 3% of the dose as 4-chloroglycolic acid anilide and 3% as 4-chlorooxalic acid anilide. After repeated intraperitoneal administration (once a week, 6 times in total), the 4-chloroglycolic acid anilide excretion was increased to 17% and the 4-chlorooxalic acid anilide excretion to 8% of the administered dose (Kiese and Lenk, 1971).

The same authors also carried out studies in pigs. After a single intraperitoneal injection of 20 mg p-chloroaniline/kg body weight, 8.6% of the dose was found in the urine as 4-chloroglycolic acid anilide. 4-Chlorooxalic acid anilide was undetectable. Similar results were obtained following the administration of a toxic dose (50 mg/kg body weight; Kiese and Lenk, 1971).

p-Chloronitrosobenzene was detected in the blood of dogs after a single intravenous injection of 25 or 100 mg p-chloroaniline/kg body weight (Kiese, 1963).

Rats were given a single intraperitoneal injection of 80 mg p-chloroaniline/kg body weight in peanut oil. After 24 h, the ether sulfate excretion in the urine was increased, whereas the glucuronide excretion was not. The sulfate conjugate of 2-amino-5-chlorophenol was reported to be the probable main metabolite (Elson et al., 1958).

Numerous in vitro studies have also been carried out on the metabolism of p-chloroaniline.

p-Chloroaniline has been shown to undergo N-hydroxylation in rat liver microsomes, with 4-chlorophenylhydroxylamine detected as a metabolite (Ping Pan et al., 1979).

Rabbit liver microsomes hydroxylated p-chloroaniline to p-aminophenol to a limited extent (Daly et al., 1968; Ichikawa et al., 1969).

Liver microsomes from phenobarbital-treated rabbits hydroxylated p-chloroaniline to 4-chlorophenylhydroxylamine and 2-hydroxy-4-chloroaniline (Lenk and Sterzl, 1981, 1984).

It was reported that N- or C-oxidation of p-chloroaniline may not only be carried out by the microsomes, but also by haemoglobin, although only to a limited extent. Its peroxidase activity appears to have an important role to play in this process (Golly and Hlavica, 1983 a, b; Corbett and Corbett, 1985).

It has also been reported that lipid peroxidation (Golly and Hlavica, 1984; Golly et al., 1984) and prostaglandin synthetase (Golly and Hlavica, 1985; Hlavica and Golly, 1985) are involved in the N-oxidation of p-chloroaniline.

In vitro, p-chloroaniline caused glutathione depletion in liver microsomes from the mouse, rat, guinea-pig and rabbit, an effect which was clearly weaker in rat liver microsomes than in microsomes from the other species (Aikawa et al., 1978).

p-Chloroaniline increased the rate of fatty acid synthesis by 3- to 4-fold at concentrations of 0.3 to 0.5 mg/ml in isolated rat liver slices (Messner et al., 1979).

7.2 Acute and subacute toxicity

Like aniline, p-chloroaniline causes methaemoglobin formation, the main symptom of which is cyanosis. According to several reports in the literature, this effect is even stronger than that caused by aniline at equimolar doses (McLean et al., 1969; Nomura, 1975; Smith et al., 1977; Birner and Neumann, 1988).

Mice (25 to 30 g) were given a single intraperitoneal injection of 0.5 mmol p-chloroaniline (hydrochloride)/kg body weight. The methaemoglobin level in the blood was 61.5 ± 1.2% after 10 min, 67.7 ± 1.5% after 30 min, 36.7 ± 4.9% after 90 min, 9.3 ± 1.6% after 150 min and 4.5 ± 0.2% after 24 h. The corresponding methaemoglobin concentrations after the same dose of aniline were 11.2 ± 1.2%, 16.6 ± 2.9%, 4.8 ± 0.4%, 1.0 ± 0.3% and 0.3 ± 0.2%. The sulfhaemoglobin level after administration of p-chloroaniline was only significantly increased after 24 h at 4.7 ± 0.6%, reaching a maximum after 48 h at 6.9 ± 1.1%. Sulfhaemoglobin was not significantly increased after administration of aniline (Nomura, 1975).

A maximum methaemoglobin level of 49.0 ± 4.5% occurred after 1 h in female Wistar rats after a single oral dose of 0.6 mmol p-chloroaniline/kg body weight (dissolved in 1,2-propanediol). After the same dose of aniline, the maximum was only 1.7 ± 0.7% and was reached after half an hour (Birner and Neumann, 1988).

Similar results were obtained in New Zealand rabbits, although the differences were not so marked. After a single intravenous injection of 0.025 mmol p-chloroaniline (hydrochloride)/kg body weight, the maximum methaemoglobin concentration was 3% after 10 min but was less than 1% after the same dose of aniline (Smith et al., 1977).

More impressive were the findings in cats, which experience has shown are particularly sensitive with regard to methaemoglobin formation. Adult cats (24 weeks old) were given 0.0625 mmol p-chloroaniline/kg body weight as a single oral dose (no further details). The average value in 5 cats was 17.1 ± 3.1% methaemoglobin after 1 hour, 33.1 ± 6.0% after 2 h, 46.4 ± 7.5% after 3 h, 53.8 ± 7.6% after 4 h and 57.8 ± 8.1% after 5 h. After 8 h, the methaemoglobin concentration had fallen to 47.8 ± 9.6%. The corresponding values after oral administration of 0.0625 mmol aniline/kg body weight were 25.4 ± 5.8%, 30.6 ± 5.3%, 28.0 ± 3.9%, 24.8 ± 4.6% and 17.5 ± 2.8% (McLean et al., 1969).

Male cats (2.8 to 4.5 kg) were given single doses of 10 mg (1 cat), 50 mg (2 cats) or 100 mg (1 cat) p-chloroaniline (≤ 99% pure)/kg body weight as a 1.5 or 10% solution in polyethylene glycol 400 by gavage. Methaemoglobin and Heinz bodies were studied 3, 7, 24 and 30 h after administration. The cat treated with 100 mg/kg body weight died after 30 h, as did 1 of the 2 cats treated with 50 mg/kg body weight. Their methaemoglobin levels were over 70%, the Heinz body counts were up to 100%. The 10 mg/kg dose led to a maximum methaemoglobin level of 28% after 3 h and a maximum Heinz body count of 39% after 7 h (Bayer, 1984).

Other studies have found that the methaemoglobin-forming potential of p-chloroaniline after oral or parenteral administration at equimolar doses was approximately equivalent to that of aniline in rats and other species (Kiese, 1974; Watanabe et al., 1976).

When p-chloroaniline was given as a single oral dose of 3 to 30 mg/animal to Wistar rats (about 15 to 150 mg/kg body weight), to one dog per dose level (about 5 to 10 mg/kg body weight) and to monkeys (30 to 60 mg/kg body weight), dose-dependent methaemoglobin formation and cyanosis occurred after 1 to 2 h. These effects disappeared after 18 to 24 h (Scott and Eccleston, 1967).

The LD_{50} values for p-chloroaniline shown in Table 1 have been reported or, in the case of older studies, can be inferred from the results given. For the rat, the LD_{50} value on oral administration was 300 to 425 mg/kg, while on dermal application it was 3200 mg/kg in one study, but only 335 mg/kg body weight in another. The dermal LD_{50} value in cats, was however ca. 250 mg/kg body weight. Signs of toxicity observed included excitation, tremors, spasms, shortness of breath and cyanosis.

After Fischer-344 rats (200 to 300 g) were given a single intraperitoneal injection of p-chloroaniline at doses of 0.4, 1 or 1.5 mmol/kg body weight, kidney damage with reduced urine volume occurred at the highest dose, with haematuria, proteinuria, increased urea nitrogen and reduced p-aminohippuric acid accumulation in the kidney cortex. This effect was, however, weaker than that of aniline (Rankin et al., 1986).

When male Wistar rats were injected once intraperitoneally with 1 to 1.5 mmol p-chloroaniline/kg body weight, there was a dose-dependent increase in the alanine aminotransferase activity and an increase in blood urea nitrogen (BUN) concentration within 24 h. The authors concluded from this that p-chloroaniline impairs the function of the liver and kidneys (Valentovic et al., 1993).

In another study 5 male Fischer-344 rats (180 to 200 g) were given a single dose of 1.0 mmol p-chloroaniline/kg body weight in olive oil by intraperitoneal injection. The urine was then collected for 24 h and the blood and kidneys were examined after 48 h. p-Chloroaniline significantly increased ($p<0.05$) the urine volume from 8.3 ml/24 h (control) to 16.8 ml/24 h, while creatinine excretion remained unaffected. In the urine, significant increases in the activities of N-acetyl-b-D-glucosaminidase at 111 mU/24 h (control 34.9 mU/24 h) and of γ-glutamyl transpeptidase at 11.3 mU/24 h (control 5.9 mU/24 h) were seen ($p < 0.01$ and $p < 0.05$, respectively). The activities of aspartate aminotransferase and alanine aminotransferase and the

Table 1. Acute toxicity of p-chloroaniline

Species	Sex	Route	LD_{50}	Reference
Rat	male	oral	380	Korte and Greim, 1981
Rat	female	oral	370	Korte and Greim, 1981
Rat	male	oral	340	Marty and Wepierre, 1979
Rat	female	oral	300	Marty and Wepierre, 1979
Rat	male	oral	340	Matrka et al., 1978
Rat	male	oral	310 (200 – 480)	Smyth et al., 1962
Rat	male/female	oral	420	Sziza and Pod-hragyai, 1957
Rat	–	oral	368	Khamuev, 1965
Rat	–	oral	425 (380 – 476)	Kondrashov, 1969 a
Mouse	male-	oral	420	Marty and Wepierre, 1979
Mouse	female-	oral	370	Marty and Wepierre, 1979
Mouse	–	oral	400	Khamuev, 1965
Mouse	–	oral	228 (198 – 262)	Kondrashov, 1969 a
Mouse	–	oral	500	Elson et al., 1948
Guinea-pig	–	oral	350	Khamuev, 1965
Rat	male/female	i.p.	420	Sziza and Pod-hragyai, 1957
Rat	male/female	dermal	3200	Sziza and Pod-hragyai, 1957
Rat	male/female	dermal	335	Korte and Greim, 1981
Rabbit	–	dermal	360	Smyth et al., 1962
Cat	–	dermal	239 (167 – 311)	Kondrashov, 1969 a
Cat	–	dermal	ca. 200 to 300	Lehmann, 1933
Cat		subcutaneous	<125	Lehmann, 1933
Rat	male	inhalation, single 4-hour exposure	2.34 (2.20 – 2.57) mg/l	Du Pont, 1981

urea nitrogen content remained unaffected in the plasma. The methaemoglobin content in the haemolysate was significantly increased ($p < 0.001$) at 4.9% (control 1.1%). Histologically, the kidney tubules were slightly swollen, but necrosis was not evident (Yoshida et al., 1989).

Mice (25 to 30 g) were given 0.5 mmol p-chloroaniline (hydrochloride)/kg body weight daily for 3 days by intraperitoneal injection. The methaemoglobin concentration was significantly increased ($p = 0.01$) at 5.0 ± 0.6%, 48 h after the last injection, as was the sulfhaemoglobin concentration at 10.0 ± 1.4% ($p = 0.01$). The effect was more severe than that of aniline in this study (Nomura, 1975).

To study the acute inhalation toxicity, groups of male Crl:CD rats (8 weeks old, 230 to 260 g) were exposed once to a p-chloroaniline vapour/aerosol mixture for 4 h (head-only exposure, 14-day observation period). The concentrations were 1.69, 1.81, 1.92, 2.10, 2.38 and 2.66 mg/l air, the respirable fraction was between 57.3 and 94.7%. All concentrations caused, among other effects, cyanosis and lethargy for up to 24 h, weight loss of 7 to 23% and clouding of the cornea, which persisted for up to 14 days. The LC_{50} value was calculated as 2.34 (2.20 to 2.57) mg/l air (Du Pont, 1981).

In an inhalation hazard test, a single 4-hour exposure to an atmosphere saturated with p-chloroaniline at room temperature was tolerated by 6 male or female albino rats. Exposure for 8 h caused deaths (Smyth et al., 1962).

The threshold concentration for increases in Heinz bodies in the peripheral blood on inhalation of p-chloroaniline for 4 h was 22.5 mg/m^3 in mice, 21.4 mg/m^3 in cats and 36 to 37 mg/m^3 in rats (Kondrashov, 1969 a).

Groups of 16 male Crl:CD rats (initial weight 230 to 267 g) inhaled 0, 2.2, 10 or 22.2 ppm p-chloroaniline (purity unspecified) 5 times weekly, 6 h/day over a period of 2 weeks. Some of the animals were observed for 14 days after the last exposure. During and immediately after exposure, the rats in the two upper dose groups showed mild to moderate cyanosis, and those in the upper dose group also showed clouding of the cornea and alopecia, which persisted until the end of the 14-day observation period, as well as reduced body weight and rattling noises on breathing. Haematologically, the predominant concentration-dependent effect was on the red blood count, with methaemoglobinaemia and subsequently secondary anaemia. These effects were reversible, or were close to control values after the 14-day observation period. Histologically, effects at all concentrations included increased extramedullary haematopoiesis in the spleen with haemosiderin storage, which was even more pronounced after the 14-day observation period. A no effect level was not attained (Du Pont, 1982).

The subacute oral toxicity of p-chloroaniline on administration by gavage for 28 days was studied in groups of 10 male and 10 female Sprague-Dawley rats (initial weights 110–112 g) in accordance with the draft OECD guideline of 31.12.79. The lowest toxic dose was < 10 mg/kg (no further details). The bone marrow, liver and spleen were shown to be target organs (Korte and Greim, 1981).

When mice were treated orally with 100 mg p-chloroaniline/kg body weight twice daily for 5 days, ca. 50% of the animals died (no further details; Elson et al., 1948).

p-Chloroaniline was administered to groups of 5 male and 5 female F344/N rats and B6C3F1 mice by gavage at doses of 0, 25, 50, 100, 200 and 400 mg/kg body weight as the hydrochloride in aqueous solution 5 times weekly (12 doses in total over a period of 16 days). All rats in the 200 and 400 mg/kg groups died within 5 days; body weight gain was reduced in the animals in the 100 mg/kg group (male rats 19%, female rats 5%). Clinical signs of toxicity included cyanotically discoloured extremities and eyes. The animals in both of the upper dose groups were lethargic and those in the 25 and 50 mg/kg groups had breathing difficulties. Post-mortems on the surviving rats revealed spleen enlargement. The mice in the upper two dose groups died, mostly within the first week; some deaths occurred in the other dose groups. However, body weight gain in the surviving animals was comparable with that of the controls. The main symptom was cyanosis. In the 4 surviving animals in the 100 mg/kg group, diffuse haemosiderosis in the Kupffer's cells of the liver and congestion of the spleen were seen (NTP, 1989).

To study the subacute toxicity of p-chloroaniline, 7 mice were given $^{1}/_{5}$ of the LD_{50} (46 mg/kg body weight) daily, 6 times/week for 5 weeks by gavage. In this way a total of 6 times the LD_{50} was administered. At the end of the study, the value for Heinz bodies was 51.4% (control 1%), the methaemoglobin content was 13.8% (control 1.6%) and the relative weight coefficient of the spleen was 11.39 g/kg (control 3.95 g/kg). These changes were significant at $p < 0.01$ (Kondrashov, 1969 a).

In another study, p-chloroaniline vapour was administered to rats by inhalation and by exposure of the shaved skin (strain, number of rats, concentrations and signs of toxicity not reported). The study was carried out in a chamber, in which either the body or the head of the rat was exposed to the vapour. The acute study involved a single 4-hour exposure and the subacute study exposure for 4 h daily, 6 times/week for a month. The threshold dose which increased the occurrence of Heinz bodies in the erythrocytes was determined as a criterion for the toxic effect of p-chloroaniline. In the acute studies, this was 22 mg/m^3 on absorption of p-chloroaniline through the skin and 36.3 mg/m^3 on inhalation exposure. When p-chloroaniline was administered at a concentration of 1 mg/m^3 for a month, the only effect shown by the exposed rats was increased aggression (Kondrashov, 1969 b).

7.3 Skin and mucous membrane effects

The primary skin irritancy of p-chloroaniline was tested in 3 male and 3 female Russian white rabbits (initial weight 2.5 to 2.8 kg) after a single 4-hour exposure to 0.5 g of the substance in accordance with the draft OECD guideline of 31.12.1979. The results were read at the end of the exposure period and after 24, 48, 72 and 96 h. In this study, p-chloroaniline was found to be non-irritant (Korte and Greim, 1981).

In other studies 100 to 900 mg p-chloroaniline/kg body weight caused inflammation of the skin of rabbits and cats, which cleared up after 15 to 20 days (no further details; Kondrashov, 1969 b).

According to early studies, p-chloroaniline had no irritant effect on the rat skin (Sziza and Podhragyai, 1957).

The eye irritancy of p-chloroaniline was tested in 6 male Russian white rabbits (initial weight 2.5 to 2.8 kg) in accordance with the draft OECD guideline of 31.12.1979. The instillation of 100 mg into the conjunctival sac of one eye caused slight reddening (grade 1) and swelling (grade 1 to 2) of the conjunctiva; the cornea and iris appeared normal (Korte and Greim, 1981).

In contrast, after instillation of 50 mg p-chloroaniline into the conjunctival sac of rabbits, severe purulent kerato-conjunctivitis occurred, which was completely healed after 5 to 10 days (no further details; Kondrashov, 1969 b).

According to other studies, 5 µl p-chloroaniline caused severe burns of the rabbit eye (no further details; Smyth et al., 1962).

An early study showed p-chloroaniline to have a slight irritant effect on the rabbit eye (Sziza and Podhragyai, 1957).

7.4 Sensitization

In guinea pigs, p-chloroaniline was moderately sensitizing (50% of the animals) in the maximization test, a very weak sensitizer (30% of the animals) in a single injection adjuvant test and did not induce sensitization (0% of the animals) in a modified Draize test. In comparison, the sensitization rate for 2,4-dinitrochlorobenzene was 100% in all three tests (Goodwin et al., 1981).

p-Chloroaniline was only a weak sensitizer in the maximization test carried out in accordance with the OEDC draft guideline of 31.12.1979 (Korte and Greim, 1981).

In a third Magnusson and Kligman maximization test in 10 Dunkin-Hartley guinea pigs (300 to 350 g), intradermal induction was carried out with a 0.3% solution, epidermal induction with a 10% solution and dermal challenge with a 2.5% solution. Of the 10 guinea-pigs, 6 reacted positively, and this was judged to indicate moderate sensitizing potential (Scholes et al., 1992; Basketter and Scholes, 1992).

According to an earlier study, p-chloroaniline showed no sensitizing effects in the Draize test on guinea pig skin (no further details; Sziza and Podhragyai, 1957).

To study the sensitizing effect of p-chloroaniline in the local lymph node assay, 25 µl at 3 different concentrations (2.5, 5 and 10%) were applied locally to the reverse of both ears in groups of 4 female CBA/Ca mice (8 to 12 weeks old) on 3 consecutive days. On day 5, the auricular lymph nodes were isolated following intravenous injection of ^{3}H-methylthymidine, and their thymidine uptake was measured. The thymidine uptake is increased on proliferation of the lymph nodes. Criteria for a positive result were considered to be a 3-fold or higher uptake relative to the controls and concentration dependency. The studies were carried out in 4 independent institutes. Overall, p-chloroaniline was judged to be negative in this test (Basketter and Scholes, 1992; Scholes et al., 1992).

7.5 Subchronic and chronic toxicity

Groups of 10 male and 10 female Wistar-SPF rats received 8, 20 or 50 mg p-chloroaniline/kg body weight/day orally for 3 months. No effects on the blood were seen at 8 or 20 mg/kg. At 50 mg/kg an increase in Heinz bodies (> 20/100 erythrocytes) and in reticulocytes (over 2%) occurred in all animals, but no fall in haemoglobin or erythrocytes or reduction in the haematocrit value were seen. Histologically, an increase in extramedullary erythropoiesis in the spleen and liver, hyperplasia of the bone marrow and haemosiderin accumulation in the kidneys were seen after 50 mg/kg. All other organs were without pathological findings. Nothing abnormal was detected in the animals at the two lower doses (Scott and Eccleston, 1967).

White rats (no further details) were given 37 mg p-chloroaniline/kg body weight (1/10 of the LD_{50} value) in sunflower oil by gavage daily for 3 months. The rats showed reduced movement and cyanosis, and effects on the blood included a reduction in erythrocytes ($p < 0.02$ to 0.001) and in the haemoglobin content ($p < 0.01$ to 0.001), reticulocytosis ($p < 0.01$ to 0.001) and an increase in polychromatic normoblasts. Furthermore, the rats showed methaemoglobinaemia and increased urobilin in the urine ($p < 0.05$ to 0.001). Post-mortem revealed a significant increase in spleen weight ($p < 0.02$ to 0.001), while histological examination showed dystrophic changes in the liver and kidneys (Khamuev, 1967).

A more recent 3-month gavage study with p-chloroaniline in rats and mice has been reported as part of the National Toxicology Program (NTP). Groups of 10 male and 10 female Fischer-344 rats received 80, 40, 20, 10 or 5 mg/kg body weight while groups of 10 male and 10 female B6C3F1 mice received 120, 60, 30, 15 or 7.5 mg/kg body weight 5 times a week for 13 weeks. 1 of the 10 female rats given 80 mg/kg died of unknown causes. Apart from this no deaths occurred and there were no differences in organ weights between the control and treated animals with the exception of the spleen weight, which was increased dose-dependently. Body weight was only reduced in the male rats in the highest dose group (80 mg/kg). Cyanosis was observed in all rats at the highest dose level. Dose-dependent methaemoglobinaemia and anaemia were seen in both species. Histologically there were effects on the spleen, kidney and liver, which were interpreted as a consequence of the methaemoglobinaemia and haemolytic anaemia. These effects were seen marginally down to the lower dose groups. A no effect level could therefore not be derived. Doses of 18, 6 and 2 mg/kg body weight for rats and 30, 10 and 3 mg/kg body weight for mice were suggested for a 2-year study (see Sect. 7.7; Chhabra et al., 1986; NTP, 1989; Chhabra et al., 1990).

p-Chloroaniline was administered to groups of 4 male and 4 female beagle dogs orally for 3 months. The doses were 5, 10 and 15 mg/kg body weight. The 5 mg/kg dose led to a decrease in haemoglobin (> 2g/100 ml) and in erythrocytes (> $1.5 \times 10^6/\mu l$) and to a reduction in the haematocrit value (> 10%) in 1/8 dogs, as well as to an increase in reticulocytes in 3/8 dogs and an increase in Heinz bodies in 2/8 dogs. These effects occurred more often after 10 and 15 mg/kg. The 10 mg/kg dose caused decreases in haemoglobin in 3/8 dogs, in erythrocytes in 2/8 dogs and in the haematocrit value in 3/8 dogs, as well as increases in reticulocytes in 7/8 dogs and in Heinz bodies in 6/8 dogs. After 15 mg/kg, 1 of the 8 dogs died as a consequence of severe haemolysis. Of the remaining animals, 4/7 showed a decrease in haemoglobin, 5/7 a decrease in erythrocytes, 6/7 a reduced haematocrit value, 7/7 an increase in reticulocytes and 6/7 an increase in Heinz bodies. Histopathologically, an increase in extramedullary erythropoiesis in the spleen and liver, hyperplasia of the bone marrow and haemosiderin accumulation in the kidneys were found. No pathological changes were detectable in any other tissues. The no effect level was thus <5 mg/kg body weight (Scott and Eccleston, 1967).

In a further study, p-chloroaniline was administered daily by gavage as a solution in sunflower oil to rats and guinea-pigs for 8 and 7 months, respectively. The doses were 0, 0.001, 0.01 and 0.1 mg/kg body weight in rats and 0, 0.05, 0.5 and 5.0 mg/kg body weight in guinea-pigs. During the study, the animals in the treated groups showed no overt differences from the controls, and there were no haematological effects. Histologically, dystrophic effects on the liver and kidney were seen in the guinea-pigs in the 5.0 and 0.5 mg/kg groups. There were no pathological effects on the blood composition of the rats and guinea-pigs, although the rats in the 0.1 mg/kg group showed an impairment of conditioned reflexes. The 0.01 and 0.001 mg/kg doses were without effect. No treatment- related effects were found in either species on histological examination (Khamuev, 1967).

In order to study the subchronic inhalation toxicity of p-chloroaniline, groups of 19 rats and 8 cats inhaled concentrations of 9.5 ± 0.7 or 1.0 ± 0.06 mg/m^3 (rats) and 6.9 ± 0.7 or 1.04 ± 0.06 mg/m^3 (cats) for 4 h/day, 6 days/week for 4 months. The parameters studied were body weight, haemoglobin, erythrocytes and Heinz bodies, as well as reticulocytes and the osmotic resistance of the erythrocytes in cats. Even after one month, the high concentration caused "considerable effects" on some of the parameters tested in the rats and the cats. In the cats, 1.04 mg/m^3 also caused an increase in Heinz bodies, which was 5.3% after 2 months, 9% after 3 months and 11.7% after 4 months ($p < 0.01$). In the rats given 1 mg/m^3, severe aggression and mild anaemia were seen after 4 months (haemoglobin 10.23 g%, controls 12.1 g%; erythrocytes 6.1 million/ µl, controls 6.59 million/ µl; $p < 0.001$ and < 0.05, respectively). The effects were not reversible in the rats after one month, while in the cats they were no longer detectable one month after the end of the study at 1.04 mg/m^3 and 2 months after the end of the study at 6.9 mg/m^3. The no effect level was < ± 1 mg/m^3 in both species (Kondrashov, 1969 a).

In a further study, rats inhaled p-chloroaniline 6 times/week at analytically controlled concentrations of 15 or 1.5 mg/m^3 for 6 months or 0.15 mg/m^3 for 3 months. Body weight gain and relative organ weights (no further details) were unaffected. At 15 mg/m^3, methaemoglobin formation occurred, reaching a maximum of 22% after 3 to 4 months, while after 6 months, the haemoglobin content was reduced (no further details), Heinz bodies were increased to 15% and a reticulocytosis of up to 10% (controls 2%) occurred. Coproporphyrin excretion was increased (no further details) in 16-hour urine, as was the level of urobilin bodies (up to 0.75 mg, controls 0.1 mg). Liver function was unaffected after 6 months. A disturbance of conditioned reflexes was detectable (no further details), which returned to normal in the fourth month. At 1.5 mg/m^3, only slight methaemoglobinaemia of 4% (controls 1.2%) occurred after 2 to 3 months. A concentration of 0.15 mg/m^3 caused no changes in the parameters studied, even the most sensitive, that is methaemoglobin, Heinz bodies, coproporphyrin and urobilin bodies. The authors considered 15 mg/m^3 to be toxic, 1.5 mg/m^3 to be the "threshold level" and 0.15 mg/m^3 as having no effect (Zvezdaj, 1970).

7.6 Genotoxicity

7.6.1 In vitro

In the Salmonella/microsome test in strains TA 98, TA 100, TA 1535, TA 1537 and TA 1538 and in G 46, TA 92 and TA 94, p-chloroaniline was negative with and without metabolic activation, up to non-cytotoxic concentrations of 1000 µg/plate (Garner and Nutman, 1977; Simmon, 1979 a; Rosenkranz and Poirier, 1979; Gilbert et al., 1980; Zimmer et al., 1980; Miyata, 1981; Thompson et al., 1983; Pai et al., 1985; Rashid et al., 1987).

Other Salmonella/microsome studies gave threshold effects or showed slight mutagenic activity, which often revealed no concentration-dependency, at concentrations of up to 1000 µg/plate (Seuferer et al., 1979; Van der Bijl et al., 1984; Dunkel et al., 1985; Mortelmans et al., 1986; Zeiger, 1990).

In a Salmonella/microsome test carried out in accordance with the draft OECD guideline of 31.12.1979, p-chloroaniline showed no mutagenic activity in strains TA 1535, TA 1537 and TA 100 with or without metabolic activation (S9-mix from Clophen A50 pretreated rat liver). The effect in TA 1538 and TA 98 was questionable (no information on concentration; Korte and Greim, 1981).

A concentration-dependent genotoxic effect occurred from 666 µg/plate in strain TA 98 after metabolic activation. Slight cytotoxicity occurred from 5000 µg/plate (Dunkel and Simmon, 1970).

In other studies, p-chloroaniline was mutagenic in TA 100 after metabolic activation, but only at concentrations close to those causing bacteriotoxicity, i.e. between about 600 and 1000 µg/plate (McGregor et al., 1984).

p-Chloroaniline had no genotoxic activity in *Escherichia coli* WP2 and WP2uvrA$^-$ or in WP2uvrA(P) and WP2uvrA at concentrations of up to 1000 µg/plate (Thompson et al., 1983; Dunkel et al., 1985; Pai et al., 1985).

In the mouse lymphoma test in L5178Y cells, p-chloroaniline increased the mutant frequency with metabolic activation by a maximum of 2.2- to 3.3-fold at concentrations of 15.7 to 24.6 µg/ml and by 2.8-fold at 15.6 to 75 µg/ml (Mitchell et al., 1988; Myhr and Caspary, 1988).

However, at higher concentrations (up to 550 µg/ml), p-chloroaniline was clearly positive in the mouse lymphoma test (L5178Y cells) with and without metabolic activation, in tests in two independent institutes (Caspary et al., 1988; Myhr et al., 1990).

In a further mouse lymphoma test, p-chloroaniline showed a dose-dependent positive effect ($p \leq 0.001$) at concentrations of 1.520 to 2.540×10^{-3} mol/l without metabolic activation. The mutation index increased with increasing dose from 0 (control) to 2.7, 3.2 and 4.3 (Wangenheim and Bolcsfoldi, 1988).

p-Chloroaniline caused no DNA strand breaks in mouse lymphoma cells at concentrations of from 0.507 to 3.04×10^{-3} mol/l without metabolic activation (not tested with metabolic activation; Garberg et al., 1988).

The effect of p-chloroaniline on the sister chromatid exchange rate (SCE) in Chinese hamster ovary (CHO) cells was also tested in two different laboratories. In one study, an increase in the exchange rate occurred without metabolic activation from 16.7 µg/ml and with activation from 900 µg/ml, while the effect in the other study was weakly positive without metabolic activation at 160 µg/ml and negative with activation up to the cytotoxic concentration of 600 µg/ml (Anderson et al., 1990).

p-Chloroaniline showed no genotoxic activity in the umu test in *Salmonella typhimurium* TA 1535/pSK1002 at concentrations of 100 to 800 mg/ml with and without metabolic activation (commercial S9-mix, no further details; Sakagami et al., 1988, 1989; Ono et al., 1992).

In the *Escherichia coli*-Pol A test with and without metabolic activation, 5 µg p-chloroaniline/ml caused DNA damage (Rosenkranz and Poirier, 1979).

p-Chloroaniline was negative in the mitotic recombination test in *Saccharomyces cerevisiae* at a concentration of 2 mg/ml with and without metabolic activation (Simmon, 1979 b).

In *Aspergillus nidulans*, 200 μg p-chloroaniline/ml increased the $meth_3$-revertant count from 3 (control) to 28 (Prasad, 1970).

In the DNA repair test (UDS test) in primary rat hepatocyte cultures, p-chloroaniline was positive at concentrations of 10^{-2} and 5×10^{-3} mg/ml. A concentration of 5×10^{-2} mg/ml was toxic (Williams et al., 1982).

In contrast, a further DNA repair test with p-chloroaniline in rat hepatocytes at concentrations of from 0.05 to 1.0 mM (in DMSO) gave negative results (Korte and Greim, 1981).

In the UDS test in cultured rat hepatocytes, 50 nmol p-chloroaniline/ml was ineffective (Thompson et al., 1983).

p-Chloroaniline was tested for clastogenicity in Chinese hamster ovary (CHO) cells in two different laboratories. In one study, 500 μg/ml was cytotoxic without metabolic activation, while with metabolic activation a positive effect was observed from 900 μg/ml. In the other study, the highest tested dose of 500 μg/ml had no effect without metabolic activation while 600 μg/ml was cytotoxic with metabolic activation (Anderson et al., 1990).

Overall, p-chloroaniline should be evaluated as being genotoxic, based on the in vitro studies.

7.6.2 In vivo

In a micronucleus test, 3 groups of 5 male and 5 female CFLP mice were given the maximum tolerated dose (as determined in preliminary studies) of 180 mg p-chloroaniline/kg body weight as a suspension in traganth as a single dose by gavage. Further groups of 5 males and 5 females served as positive (40 mg cyclophosphamide/kg body weight) and negative (aqueous traganth preparation) controls. Femoral bone marrow was examined for micronuclei 24, 48 and 72 h after administration (1000 polychromatic erythrocytes/mouse). In this study, p-chloroaniline caused no significant increase in micronuclei and was evaluated as negative (TLL, 1986/87).

p-Chloroaniline was genotoxic in the somatic mutation and recombination test after feeding 3-day-old trans-heterozygous *Drosophila melanogaster* larvae 7.84 mM (equivalent to 1 mg/ml) for 6 h. There was an increase in single and double spots on the wings (Graf et al., 1990).

7.7 Carcinogenicity

Several long-term carcinogenicity studies have been carried out with p-chloroaniline in rats and mice:

In the first study, groups of 50 male and 50 female Fischer-344 rats and 50 male and 50 female B6C3F1 mice were given 250 or 500 mg p-chloroaniline/kg diet (rats) or 2500 or 5000 mg/kg diet (mice) for 78 weeks. The animals were then observed for periods of 24 weeks in the rats and 13 weeks in the mice. In the high dose female rats, body weight gain was slightly inhibited from week 40 of the study and in the high dose male rats, survival was somewhat reduced. However, an adequate number of animals survived for the entire study (38/50 in the high dose group, 46/50 in the low dose group and 18/20 controls). In male, but not female, rats given the top dose, 6 fibromas and 4 sarcomas of the spleen were found (1 fibrosarcoma, 1 haemangiosarcoma, 1 osteosarcoma and 1 capsular carcinoma). These tumours are extremely rare in Fischer-344 rats (historical control value in males: 0/360). In male and female mice, a clearly lower body weight gain was evident in both dose groups. While the controls attained final weights of 42 g (males) and 37 g (females), the final weights of the treated mice (both sexes) were 8 to 10 g lower. Survival, however, was equivalent to that in the controls. Increased numbers of haemangiomas and haemangiosarcomas of the spleen and kidneys were found in mice of both sexes, but the increases were not statistically significant. In addition, non-tumorous proliferation and chronic inflammation of the spleen were observed in rats and mice as chronic toxic effects. This was not significant in either species. The rarity of the spleen tumours seen in the male rats and the fact that the spleen was a target organ of carcinogenic activity in the mice, raises the suspicion that these tumours were caused by p-chloroaniline. However, the authors adopted the view that the study had not produced sufficient proof of the carcinogenicity of p-chloroaniline (NCI, 1979).

This study was later repeated by gavage, as it was suggested that p-chloroaniline was not stable in the feed, and the animals could have consumed less than the reported doses.

Groups of 50 male and 50 female Fischer-344 rats were given 2, 6 or 18 mg (99.1% pure) p-chloroaniline/kg body weight as the hydrochloride 5 times weekly for 103 weeks by gavage. Body weight gain was within the range of that of the controls, while survival of the males in the low and intermediate dose groups and of the females in the low and high dose groups was significantly increased relative to the controls. This was attributed to the lower incidence of monocytic leukaemia in the treated animals. The incidence of spleen sarcomas was significantly higher in the males in the top dose group (fibrosarcomas, osteosarcomas, haemangiosarcomas; controls 0/49, low dose 1/50, intermediate dose 3/50, high dose 38/50). The tumours had numerous metastases. In the females, only 1 fibrosarcoma and 1 osteosarcoma of the spleen were seen in the intermediate and top dose groups, respectively (historical control values for spleen sarcomas: males 1/298, females 0/297; for haemangiosarcomas of the spleen: males 0/300, females 1/297). In addition, the incidence of fibrosis of the spleen was increased in males in all groups and females in the high dose group (males: controls 3/49, low dose 11/50, intermediate dose 12/50, high dose 41/50; females: controls 1/50, low dose 2/50, intermediate dose 3/50, high dose 42/50). There was also significant fatty metaplasia in the spleen in the rats at the high dose level (males: controls 0/49, low dose 0/50, intermediate dose 0/50, high dose 24/50; females: controls 0/50, low dose 0/50, intermediate dose 0/50, high dose 11/50). There was an increased incidence of hyperplasia of the adrenal medulla in the high dose females (controls 4/50, low dose 4/50, intermediate dose 7/50, high dose 24/50). The incidence of phaeochromocytomas of the adrenal gland was marginally increased in both sexes at the top dose level (males: controls 13/49, low dose 14/48, intermediate dose 15/48, high dose 26/49; females: controls 2/50, low dose 3/50, intermediate dose 1/50, high dose 6/50). Monocytic leukaemia occurred in the treated rats to a lesser extent than in the controls (males: controls 21/49, low dose 3/50, intermediate dose 2/50, high dose 3/50; females: controls 10/50, low dose 2/50, intermediate dose 1/50, high dose 1/50). Histologically, treatment-related non-neoplastic effects were seen in the bone marrow, spleen and liver (bone marrow hyperplasia, haemosiderosis of the liver, fibrosis of the spleen), and were judged to be a consequence of the general effect of p-chloroaniline on the haematopoietic system. Overall, according to the authors, this study gave clear evidence of carcinogenic activity in male rats and equivocal evidence of carcinogenic activity in female rats (NTP, 1989; Chhabra et al., 1991).

In an analogous study, groups of 50 male and 50 female B6C3F1 mice received 3, 10 or 30 mg (99.1% pure) p-chloroaniline/kg body weight as the hydrochloride 5 times weekly for 103 weeks by gavage. Body weight gain of mice in the top dose group corresponded to that of the controls and the survival rate for male mice in the intermediate dose group was markedly lower than that of the controls after 99 weeks (males: controls 43/50, low dose 36/50, intermediate dose 29/50, high dose 35/50; females: controls 39/50, low dose 42/50, intermediate dose 44/50, high dose 41/50). In the males, there were increased incidences of hepatocellular adenomas or carcinomas (controls 11/50, low dose 21/49, intermediate dose 20/50, high dose 21/50) and haemangiosarcomas (controls 4/50, low dose 4/49, intermediate dose 1/50, high dose 10/50), although no dose-dependency was evident. In the mice at the high dose level, the liver carcinomas had very often metastasized into the lung. No treatment-related tumours occurred in the female mice. The incidence of lymphomas was reduced in both sexes (males: controls 10/50, low dose 3/49, intermediate dose 9/50, high dose 3/50; females: controls 19/50, low dose 12/50, intermediate dose 5/50, high dose 10/50). On the whole, in the opinion of the authors, p-chloroaniline showed some evidence of carcinogenic activity in male mice and no evidence of carcinogenic activity in female mice in this study (NTP, 1989; Chhabra et al., 1991).

The ability of p-chloroaniline to induce lung adenomas in strain A/St mice, which are particularly susceptible to this type of tumour, was studied. The adenomas begin spontaneously at the age of 3 to 4 months, and reach nearly 100% after 24 months. Groups of 10 male and 10 female mice were given intraperitoneal injections of 57.5 or 25 mg p-chloroaniline/kg body weight in tricaprylin (0.1 ml/mouse) 3 times weekly for 8 weeks and then examined for lung adenomas 16 weeks after the last injection. At the end of the study, 4/10 males and all 10 females in the high dose group, and all 10 animals of either sex in the low dose group still survived. No lung adenomas were detected (Maronpot et al., 1986).

A transformation test was carried out in rat embryo cells ($2FR_450$) infected with Rauscher leukaemia virus. An increase in transformed cells of 146% over the controls was caused by 19.0 µg/5.2 x 10^4 cells. This effect was evaluated as positive (Traul et al., 1981).

p-Chloroaniline was also reported to be positive in a cell transformation test in hamster embryo cells. The concentration range tested was 0.01 to 100 µg/ml and the effective concentrations were reported to be 0.01, 1, 10 and 100 µg/ml (Pienta and Kawalek, 1981). However, in an earlier publication by one of the authors, a negative result was obtained in the same test with the same doses (Pienta et al., 1977).

Results for the cell transformation test in C3H/10T1/2 mouse embryo cells are available from two independent laboratories. One laboratory tested p-chloroaniline at concentrations of 0.8, 4, 20 and 100 µg/ml, the other at 30, 100, 300 and 1000 µg/ml. 1000 µg/ml was cytotoxic, and the results at the other concentrations were positive (Dunkel et al., 1988).

A cell transformation test in BALB/C3T3 cells was negative (no further details; NTP, 1983).

7.8 Reproductive toxicity

No information available.

7.9 Effects on the immune system

Changes in immunobiological reactions were observed in the guinea pigs at the highest dose level in the study described in Sect. 7.5 (0, 0.05, 0.5, 5.0 mg/kg body weight orally for 7 months). The phagocytic activity of the leukocytes initially increased significantly ($p<0.01$) during the last month of the study and then was significantly reduced. Dose levels of 0.5 and 0.05 mg/kg were without effect (Khamuev, 1967).

7.10 Neurotoxicity

No information available.

7.11 Other effects

p-Chloroaniline (127.57 to 510.28 mg/µl) inhibited DNA and RNA synthesis in *Bacillus subtilis* in vitro (concentration-dependently) by a maximum of 50% and increased the replication period of DNA by inactivation of DNA polymerase I (Ogawa et al., 1991).

A concentration-dependent inhibition of mitotic activity occurred in human lymphocyte cultures at 20, 100 and 500 µg p-chloroaniline/ml (Vassileva and Kazakova, 1978).

When p-chloroaniline was administered by intraperitoneal injection to rabbits as a single dose of 500 mg/kg body weight, a decrease in blood pressure and a reduction in myocardial glycogen content occurred (Labunskii, 1972).

Furthermore, it has been reported that p-chloroaniline is able to form adducts and conjugates with haemoglobin (Lewalter and Korallus, 1985).

Of 13 different monocyclic aromatic amines administered orally to female Wistar rats at 0.6 mmol/kg body weight, p-chloroaniline had by far the highest haemoglobin-binding index at 569 mmol/mol after 24 h. Analogous effects were found in B6C3F1 mice, although a haemoglobin-binding index of 132 following a dose of 1 mmol p-chloroaniline/kg body weight was lower than that in the rat. Again this value was far higher than that of any of the other aromatic amines studied (Birner and Neumann, 1987, 1988).

In a further study by the same group, the differential binding of labelled p-chloroaniline in the erythrocytes and the plasma of rats was reported. After a single oral dose of 14 µmol/kg body weight, the haemoglobin-binding index (HBI) was 162 and the plasma-protein-binding index (PPBI) was 5.5 after 24 h. Of this, 93 and 24%, respectively could be hydrolyzed. The ratio of HBI to PPBI was 29.3. p-Chloroaniline also had the highest HBI of several aromatic amines in this study (Neumann et al., 1993).

8. Experience in humans

The dermal absorption of p-chloroaniline by human skin was tested in vitro with the radiolabelled material (^{14}C; 3.5 mCi/mmol) in various carriers. In these studies, abdominal skin sections from surgical operations were used. Concentrations of 26.5 µg/g polyethylene glycol 400, 26.5 µg/g vaseline plus polyethylene glycol 400 and 19.8 µg/g liquid soap plus polyethylene glycol 400 were studied. The study was carried out with untreated and scarified pieces of skin and involved 24-hour contact. In normal skin, p-chloroaniline absorption was highest with vaseline and polyethylene glycol 400 (16.4 ng/cm^2), followed by liquid soap plus polyethylene glycol 400 (14.3 ng/cm^2), and polyethylene glycol 400 (1.7 ng/cm^2). Absorption by damaged skin decreased in the order polyethylene glycol 400 (148 ng/cm^2) > liquid soap plus polyethylene glycol 400 (56.2 ng/cm^2) > vaseline and polyethylene glycol 400 (43.1 ng/cm^2). These studies showed that the absorption of p-chloroaniline through intact skin was only slight, but was markedly greater through damaged skin. Moreover, the absorption-inhibiting effect of polyethylene glycol 400 on the intact skin was very marked (Marty and Wepierre, 1979).

In an acute case of p-chloroaniline poisoning (no detailed data on case history), urinary metabolites were studied in the patient during his stay in hospital (by gas chromatography, mass spectrometry, HPLC and colorimetric tests for diazo-positive metabolites). The main metabolites were conjugated p-chloroaniline and conjugated 2-amino-5-chlorophenol, while the amounts of 2,4-dichloroaniline and free p-chloroaniline were low. The half-lives for p-chloroaniline, 2-amino-5-chlorophenol and 2,4-dichloroaniline were 2.4, 1.7 and 1.7 h, respectively during the rapid phase of elimination and 4.5, 3.3 and 3.8 h, respectively during the slow phase. About 62% of the total excretion was conjugated or free p-chloroaniline, about 36% was 2-amino-5-chlorophenol and about 1% was 2,4-dichloroaniline. The total balance was thus about 99%. The authors recommended the measurement of p-chloroaniline and its conjugates and of 2-amino-5-chlorophenol in the urine by HPLC for biological monitoring of p-chloroaniline exposure (Yoshida et al., 1991, 1992 a, b).

The acute toxic effect (methaemoglobin formation) of p-chloroaniline in humans is more severe than that of aniline (Linch, 1974; Gosselin et al., 1984). Alcohol can lead to an increase in the severity of the toxic symptoms (Kühn-Birett, 1980).

In 1926, a fatal case of p-chloroaniline poisoning was reported in the annual report of a state industrial doctor. During inspection of a blocked duct, a worker was sprayed in the face and on the upper part of his clothing with hot p-chloroaniline. On admission to the sick room, the patient was deeply unconscious, his breathing was laboured, and the face and mucous membranes of the lips were grey-blue. The venous blood was viscous and chocolate-brown, methaemoglobin bands were clearly visible spectroscopically. Death occurred in hospital 32 h after the accident (Betke, 1926).

In a serious case of acute p-chloroaniline poisoning, methaemoglobinaemia of up to 43% and temporary ECG changes occurred in a 32-year-old man. The ECG effects corresponded with the severity of the clinical symptoms and were regarded as being a consequence of temporary anoxia. The patient recovered and was discharged (Scotti and Tomasini, 1966).

A further case of p-chloroaniline poisoning was reported from France. A 34-year-old employee was working with the powdered substance, which contaminated his forearms and hands. After 3 h of exposure the worker developed symptoms of weakness, somnolence, headache, dizziness, tinnitus, dyspnoea and retrosternal pain. The doctor diagnosed cyanosis, which was particularly marked on the nails and lips. A methaemoglobin level of 70% was determined in the clinic. After intravenous administration of vitamin C and particularly of methylene blue, the methaemoglobin level fell initially to 13% and then to 5% and, having recovered, the patient was discharged (Faivre et al., 1971).

p-Chloroaniline concentrations of 44 mg/m^3 air are said to lead to severe signs of toxicity in man after exposure for one minute, while 22 mg/m^3 leads to signs of illness after longer periods of inhalation (no further details; Goldblatt, 1955).

The health of 90 workers who were employed in the manufacture of monuron and who, in addition to p-chloroaniline, were also exposed to chlorobenzene, phosgene, dimethylamine and p-chlorophenyl isocyanate, was studied for 6 months. The concentration of p-chloroaniline at the workplace was 0.2 to 2.0 mg/m^3 and in most cases was within tolerable limits. At the end of the shift, 14 workers showed increased methaemoglobin concentrations, increased Heinz bodies and a tendency towards a reduced haemoglobin level. In 4 cases, there was mild p-chloroaniline intoxication with headaches, nausea, cyanosis, reduced haemoglobin and erythrocyte levels, methaemoglobin formation and an increased Heinz body count (Levina et al., 1966).

In a production plant, p-chloroaniline concentrations of 63 (37 to 89) and 58 (46 to 70) mg/m^3 were measured. Cyanosis was a common finding in the 6 people employed there. The methaemoglobin level, at 1.6 to 2.8 g/100 ml blood, was higher than that in workers in other plants, as was the sulfhaemoglobin level, at 0.5 g/100 ml blood. During the 4-week production period, 2 of the 6 workers developed anaemia, and one had to stop work due to acute intoxication. In the authors' opinion, dermal absorption was implicated in this intoxication to a considerable extent, as well as inhalation (Pacséri et al., 1958).

The haemoglobin, methaemoglobin and reticulocyte levels were studied in 14 workers in a plant manufacturing p-chloroaniline from p-chloronitrobenzene. They worked 3 12-hour shifts or 5 8-hour shifts per week. The concentrations of p-chloroaniline and p-chloronitrobenzene in the air were simultaneously monitored (although the results were not presented). For comparison, 21 people from other areas were used. There was a significant fall in haemoglobin in the exposed subjects ($p < 0.04$). Following exposure, the methaemoglobin level was significantly increased ($p < 0.02$), but was not correlated with the results of the air analysis (Monsanto, 1986).

Methaemoglobin concentrations of 14.5, 43.5 and 34% were measured in 3 premature babies (25 to 27 weeks) that were treated in a new model incubator in a children's hospital. After the premature babies with the highest levels were successfully treated with intravenous methylene blue, all 3 were placed in another type of incubator and made complete recoveries. A check on the new equipment revealed that chlorhexidine gluconate had accidentally been used in place of distilled water as an air moistening agent, and had evaporated at 100 °C in the new equipment. In this way, p-chloroaniline (one of the starting substances in the synthesis of chlorhexidine gluconate) had been liberated, and had caused the methaemoglobinaemia (van der Vorst et al., 1990).

The ability of aromatic amines to form conjugates and adducts with haemoglobin has been detected in man. The people involved had been occupationally exposed to p-chloroaniline or aniline and had suffered acute poisonings. Conjugate formation due to p-chloroaniline was correlated quantitatively with the methaemoglobin level and was detectable not only 30 min after intoxication but also for up to 7 days thereafter. In the urine, on the other hand, p-chloroaniline and/or its metabolites were detectable for up to 16 h only. Analogous findings resulted with aniline. The authors suggested this conjugate formation could be used for biological monitoring of exposure (Lewalter and Korallus, 1985).

9. Threshold limit values

p-Chloroaniline is classified as carcinogen category 2, substances that should be regarded as carcinogenic in man (TRGS 905, 1994).

References

Aikawa, K., Satoh, T., Kobayashi, K., Kitagawa, H.
Glutathione depletion by aniline analogs in vitro associated with liver microsomal cytochrome P-450
Japan. J. Pharmacol., 28, 699–705 (1978)

Anderson, B.E., Zeiger, E., Shelby, M.D., Resnick, M.A., Gulati, D.K., Ivett, J.L., Loveday, K.S.
Chromosome aberration and sister chromatid exchange test results with 42 chemicals
Environ. Mol. Mutagen., 16, Suppl. 18, 55–137 (1990)

Basketter, D.A., Scholes, E.W.
Comparison of the local lymph node assay with the guinea-pig maximization test for the detection of a range of contact allergens
Food Chem. Toxicol., 30 (1), 65–69 (1992)

Bayer AG, Institut für Toxikologie
p-Chloranilin, Untersuchungen zur akuten oralen Toxizität an der Katze. Einfluß auf Met-Hämoglobingehalt und Zahl der Heinz-Innenkörper im peripheren Blut
Unpublished report (1984)

Betke
Jahresbericht des Gewerbemedizinalrates für den Aufsichtsbezirk Wiesbaden über das Geschäftsjahr 1922
Veröffentlichungen aus dem Gebiet der Medizin, Vol. 21, 212 (578) – 258 (624) (1926)

Bijl, Van der, P., Gelderblom, W.C.A., Thiel, P.G.
On the mutagenicity of parachloroaniline, a breakdown product of chlorhexidine
J. Dent. Ass. S. Afr., 39, 535–537 (1984)

Birner, G., Neumann, H.G.
Biomonitoring of aromatic amines, binding of substituted anilines to rat hemoglobin
Naunyn Schmiedebergs Arch. Pharmacol., 335, R. 19 (1987)

Birner, G., Neumann, H.G.
Biomonitoring of aromatic amines II: Hemoglobin binding of some monocyclic aromatic amines
Arch. Toxicol., 62, 110–115 (1988)

Bray, H.G., James, S.P., Thorpe, W.V.
The metabolism of the monochloronitrobenzenes in the rabbit
Biochem. J., 64, 38–44 (1956)

Caspary, W.J., Daston, D.S., Myhr, B.C., Mitchell, A.D., Rudd, C.J., Lee, P.S.
Evaluation of the L5178Y mouse lymphoma cell mutagenesis assay: interlaboratory reproducibility and assessment
Environ. Mol. Mutagen., 12, Suppl. 13, 195–229 (1988)

Chhabra, R.S., Gerken, D.K., Wilson, F.D., Peters, A.C.
p-Chloroaniline subchronic toxicity studies in rats and mice
Toxicol. Lett., 31, 244 (1986)

Chhabra, R.S., Thompson, M., Elwell, M.R., Gerken, D.K.
Toxicity of p-chloroaniline in rats and mice
Food Chem. Toxicol., 28 (10), 717–722 (1990)

Chhabra, R.S., Huff, J.E., Haseman, J.K., Elwell, M.R., Peters, A.C.
Carcinogenicity of p-chloroaniline in rats and mice
Food Chem. Toxicol., 29 (2), 119–124 (1991)

Corbett, M.D., Corbett, B.R.
Enzymatic N-oxidation of 4-nitroaniline
Biochem. Arch., 1, 115–120 (1985)

Daly, J.W., Guroff, G., Udenfriend, S., Witkop, B.
Hydroxylation of alkyl and halogen substituted anilines and aceta-

nilides by microsomal hydroxylases
Biochem. Pharmacol., 17, 31–36 (1968)

Dunkel, V.C., Simmon, V.F.
Mutagenic activity of chemicals previously tested for carcinogenicity in the National Cancer Institute bioassay program
IARC-Sci. Publ., 27, 283–302 (1980)

Dunkel, V.C., Zeiger, E., Brusick, D., McCoy, E., McGregor, D., Mortelmans, K., Rosenkranz, H.S., Simmon, V.F.
Reproducibility of microbial mutagenicity assays: II. Testing of carcinogens and noncarcinogens in *Salmonella typhimurium* and *Escherichia coli*
Environ. Mutagen., 7, Suppl. 5, 1–248 (1985)

Dunkel, V.C., Schechtman, L.M., Tu, A.S., Sivak, A., Lubet, R.A., Cameron, T.P.
Interlaboratory evaluation of the C3H/10T1/2 cell transformation assay
Environ. Mol. Mutagen., 12, 21–31 (1988)

Du Pont (E.I. du Pont de Nemours and Co., Newark, Delaware)
Benzenamine, 4-chloro, inhalation median lethal concentration (LC_{50})
Report no. 14050 (1981)
NTIS/OTS 84003A

Du Pont (E.I. du Pont de Nemours and Co., Newark, Delaware)
Benzenamine, 4-chloro, subacute inhalation toxicity study of p-chloroaniline in rats
Report, Haskell no. 14050 (1982)
NTIS/OTS 84003A

Ehlhardt, W.J., Howbert, J.J.
Metabolism and disposition of p-chloroaniline in rat, mouse, and monkey
Drug Metabol. Dispos., 19 (2), 366–369 (1991)

Elson, L.A., Spinks, A., Tottey, M.M.
Some pharmacological properties of p-chloroaniline
Biochem. J., 42, X/III (1948)

Elson, L.A., Goulden, F., Warren, F.L.
The metabolism of aromatic amines in relation to carcinogenesis
Br. J. Cancer, 12, 108–115 (1958)

Faivre, M., Armand, J., Évreux, J.C., Duverneuil, G., Colin, C.
Méthémoglobinémie toxique par des dérivés de l'aniline: parachloroaniline et paratoluidine
Arch. Mal. Prof., 32, 575–577 (1971)

Garberg, P., Åkerblom, E.L., Bolcsfoldi, G.
Evaluation of a genotoxicity test measuring DNA-strand breaks in mouse lymphoma cells by alkaline unwinding and hydroxyapatite elution
Mutat. Res., 203, 155–176 (1988)

Garner, R.C., Nutman, C.A.
Testing of some azo dyes and their reduction products for mutagenicity using Salmonella typhimurium TA 1538
Mutat. Res., 44, 9–19 (1977)

Gilbert, P., Saint-Ruf, G., Poncelet, F., Mercier, M.
Genetic effects of chlorinated anilines and azobenzenes on Salmonella typhimurium
Arch. Environ. Contam. Toxicol., 9, 533–541 (1980)

Goldblatt, M.W.
Research in industrial health in the chemical industry
Br. J. Ind. Med., 12, 1–20 (1955)

Golly, I., Hlavica, P.
Mechanisms of extrahepatic bioactivation of aromatic amines: the role of hemoglobin in the N-oxidation of 4-chloroaniline
In: Rydström, J., Montelius, J., Bengtsson, M. (eds.)
Extrahepatic drug metabolism and chemical carcinogenesis, 235–236
Elsevier Sciences Publishers B.V. (1983 a)

Golly, I., Hlavica, P.
The role of hemoglobin in the N-oxidation of 4-chloroaniline
Biochem. Biophys. Acta, 760, 69–76 (1983 b)

Golly, I., Hlavica, P.
The role of lipid peroxidation in the N-oxidation of 4-chloroaniline
Naunyn Schmiedebergs Arch. Pharmacol., 325, R 15 (1984)

Golly, I., Hlavica, P.
N-Oxidation of 4-chloroaniline by prostaglandin synthase. Redox cycling of radical intermediate(s)
Biochem. J., 226, 803–809 (1985)

Golly, I., Hlavica, P., Wolf, J.
The role of lipid peroxidation in the N-oxidation of 4-chloroaniline
Biochem. J., 224 (2), 415–421 (1984)

Goodwin, B.F.J., Crevel, R.W.R., Johnson, A.W.
A comparison of three guinea-pig sensitization procedures for the detection of 19 reported human contact sensitizers
Contact Dermatitis, 7, 248–258 (1981)

Gosselin, R.E., Smith, R.P., Hodge, H.C., Braddock, J.E.
Clinical toxicology of commercial products. 5th ed., II-208
Williams and Wilkins, Baltimore/London (1984)

Graf, U., Hall, C.B., van Schaik, N.
On the use of excision repair defective cells in the wing somatic mutation and recombination test in *Drosophila melanogaster*
Environ. Mol. Mutagen., 16, 225–237 (1990)

Hlavica, P., Golly, I.
N-Oxidation of 4-chloroaniline by the microsomal prostaglandin synthetase system: redox cycling of anitroxide intermediate
In: Gorrod, J.W.W., Damani, L.A. (eds.)
Biological oxidation of nitrogen in organic molecules. Chemistry, toxicology and pharmacology, 320–326
VCH Verlagsgesellschaft, Weinheim (1985)

Hoechst AG
Product data sheet "p-Chloranilin" (undated)

Ichikawa, Y., Yamano, T., Fujishima, H.
Relationship between the interconversion of cytochrome P-450 and P-420 and its activities in hydroxylations and demethylations by P-450 oxidase systems
Biochim. Biophys. Acta, 171, 32–46 (1969)

Ismail, R., Hockwin, O., Korte, F., Klein, W.
Environmental chemical permeation of bovine ocular lens capsule
Chemosphere, 2, 145–150 (1976)

Ismail, R.M., Salminen, L., Korte, F.
Distribution of topically applied environmental chemicals in the rabbit eye
Chemosphere, 12, 797–802 (1977)

Jagow, von, R., Kiese, M., Renner, G., Wiedemann, I.
Auscheidung von N-Hydroxylierungsprodukten aromatischer Amine in Harn und Galle
Naunyn Schmiedebergs Arch. Exp. Pathol. Pharmakol., 253, 47+48 (1966)

Khamuev, G.D.
Chlorinated anilines and their sanitary-toxicological characteristics
Khim. Faktory Vnesh. Sredy Ikh Gig. Znach., 108–110 (1965)

Khamuev, G.D.
The maximum permissible concentration of p-chloroaniline and m-chloroaniline in water bodies
Hyg. Sanit., 32, 15–21 (1967)

Kiese, M.
The effect of certain substituents upon the N-oxidation of aniline in vivo
Naunyn Schmiedebergs Arch. Exp. Pathol. Pharmakol., 244, 387±404 (1963)

Kiese, M.
Methemoglobinemia: a comprehensive treatise
CRC-Press, Cleveland/Ohio (1974)

Kiese, M., Lenk, W.
Metabolites of 4-chloroaniline and chloroacetanilides produced by rabbits and pigs
Biochem. Pharmacol., 20, 379–391 (1971)

Kondrashov, V.A.
Die Toxikologie von Chloranilinen und Anilin (German translation from the Russian)
Tr. Gos. Inst. Prikl. Khim., 62, 209–214 (1969 a)

Kondrashov, V.A.
Der toxische Einflub der Dämpfe von Chloranilinen und Anilin auf den Organismus bei der Einwirkung auf die unverletzte Haut (German translation from the Russian)
Gig. Tr. Prof. Zabol., 13, issue 5, 29–32 (1969 b)

Korte, F., Greim, H.
In: Umweltforschungsplan des Bundesministeriums des Innern Überprüfung der Durchführbarkeit von Prüfungsvorschriften und der Aussagekraft der Grundprüfung des E. Chem. G.
On behalf of the German Federal Office of the Environment (Umweltbundesamt) (1981)

Kühn-Birett
Merkblätter Gefährliche Arbeitsstoffe
Verlag Moderne Industrie (1980)

Labunskii, V.V.
Effect in experiments of aromatic nitro, chloro, and amino compounds on the cardiovascular system
Farmakol. Toksikol. (Kiew), 7, 156–159 (1972)
Cited in: Chem. Abstr., 79, 25, 13556s (1973)

Lawrence, F.R., Marshall, W.J.
Aniline

In: Ullmann's encyclopedia of industrial chemistry
5th ed., vol. A2, p. 303–312
VCH Verlagsgesellschaft mbH, Weinheim (1985)

Lehmann, K.B.
Studien über die Wirkung der Chloraniline und Chlortoluidine und des salzsauren 5-Chlor-2-Toluidins
Arch. Hyg. Bakteriol., 110, 12–31 (1933)

Lenk, W., Sterzl, H.
The effect of an additional Cl-atom in 3′-position on the biotransformation in vitro of 4-chloropropionanilide and 4-chloroaniline and on the formation of ferrihemoglobin
Naunyn Schmiedebergs Arch. Pharmacol., 316, R 9 (1981)

Lenk, W., Sterzl, H.
Peroxidase activity of oxyhaemoglobin in vitro
Xenobiotica, 14 (7), 581–588 (1984)

Levina, M.M., Kurando, T.B., Belyakov, A.A., Smirnova, V.G., Odlyzhko, S.L.
Questions of occupational hygiene and the status of employee health in the production of monuron
Gig. Tr. Prof. Zabol., 10, issue 11, 54–56 (1966)
English abstract: NIOSH-00136320

Lewalter, J., Korallus, U.
Blood protein conjugates and acetylation of aromatic amines. New findings on biological monitoring
Int. Arch. Occup. Environ. Health, 56, 179–196 (1985)

Linch, A.L.
Biological monitoring for industrial exposure to cyanogenic aromatic nitro and amino compounds
Am. Ind. Hyg. Assoc. J., 35, 426–432 (1974)

Maronpot, R.R., Shimkin, M.B., Witschi, H.P., Smith, L.H., Cline, J.M.

Strain A mouse pulmonary tumor test results for chemicals previously tested in the National Cancer Institute carcinogenicity tests
J. Natl. Cancer Inst., 76 (6), 1101–1112 (1986)

Marty, J.P., Wepierre, J.
Evaluation de la toxicité d'agents d'activité cosmétique cas du trichlorocarbanilide
Labo-Pharma-Problèmes et Techniques, 27 (286), 306–310 (1979)

Matrka, M., Rambousek, V., Zvèrina, Z.
Toxicity of p-substituted derivatives of aniline in experimental rats
Ceskoslovenská Hygiena, 23 (4), 168–172 (1978)

McGregor, D., Prentice, R.D., McConville, M., Lee, Y.J., Caspary, W.J.
Reduced mutant yield at high doses in the Salmonella/activation assay: the cause is not always toxicity
Environ. Mutagen., 6, 545–557 (1984)

McLean, S., Starmer, G.A., Thomas, J.
Methaemoglobin formation by aromatic amines
Pharm. Pharmacol., 21, 441–450 (1969)

Messner, B., Berndt, J., Baum, I.
Changes in fatty acid and cholesterol synthesis in rat liver slices due to aromatic amines formed during the degradation of some herbicides
Bull. Environ. Contam. Toxicol., 21, 831–836 (1979)

Mitchell, A.D., Rudd, C.J., Caspary, W.J.
Evaluation of the L5178Y mouse lymphoma cell mutagenesis assay: intralaboratory results for sixty-three coded chemicals tested at SRI International
Environ. Mol. Mutagen., 12, Suppl. 13, 37–101 (1988)

Miyata, R., Nohmi, T., Yoshikawa, K., Ishidate, M., jr.
Metabolic activation of p-nitrotoluene and trichloroethylene by rat-liver S9 or mouse-liver S9 fractions in *Salmonella typhimurium* strains
Bull. Natl. Inst. Hyg. Sci. (Tokyo), 99, 60–65 (1981)

Monsanto Company, Department of Medicine and Environmental Health, St. Louis, Missouri, USA
Evaluation of blood methemoglobin levels for workers involved in production of parachloroaniline
Report no. MC 86–9055 (1986)
NTIS/OTS 0510320

Mortelmans, K., Haworth, S., Lawlor, T., Speck, W., Tainer, B., Zeiger, E.
Salmonella mutagenicity tests: II. Results from the testing of 270 chemicals
Environ. Mutagen., 8, Suppl. 7, 1–119 (1986)

Myhr, B.C., Caspary, W.J.
Evaluation of the L5178Y mouse lymphoma cell mutagenesis assay: intralaboratory results for sixty-three coded chemicals tested at Litton Bionetics, Inc.
Environ. Mol. Mutagen., 12, Suppl. 13, 103–194 (1988)

Myhr, B., McGregor, D., Bowers, L., Riach, C., Brown, A.G., Edwards, I., McBride, D., Martin, R., Caspary, W.J.
L5178Y mouse lymphoma cell mutation assay results with 41 compounds
Environ. Mol. Mutagen., 16, Suppl. 18, 138–167 (1990)

NCI (National Cancer Institute)
Bioassay of p-chloroaniline for possible carcinogenicity
Technical Report Series No. 189 (1979)

NTP (National Toxicology Program)
NTP Technical Bulletin No. 9 (1983)

NTP (National Toxicology Program)
Toxicology and carcinogenesis studies of para-chloroaniline hydrochloride in F344/N rats and B6C3F1 mice
Technical Report No. 351 (1989)

Neumann, H.G., Birner, G., Kowallik, P., Schütze, D., Zwirner-Baier, I.
Hemoglobin adducts of N-substituted aryl compounds in exposure control and risk assessment
Environ. Health Perspect., 99, 65–69 (1993)

Nomura, A.
Studies on sulfhemoglobin formation by various drugs
Folia Pharmacol. Japon., 71, 351–365 (1975)

Ogawa, T., Hirose, Y., Yatome, C.
Effects of monochlorophenols and p-chloroaniline on nucleic acid synthesis in microbial growth process
Bull. Environ. Contam. Toxicol., 47, 8–14 (1991)

Ono, Y., Somiya, I., Kawaguchi, T.
Genotoxic evaluation on aromatic organochlorine compounds by using umu test
Water Sci. Technol., 26 (1–2), 61–69 (1992)

Pacséri, I., Magos, L., Bátskor, I.A.
Threshold and toxic limits of some amino and nitro compounds
Arch. Ind. Health, 18, 1–8 (1958)

Pai, V., Bloomfield, S.F., Gorrod, J.W.
Mutagenicity of N-hydroxylamines and N-hydroxycarbamates towards strains of *Escherichia coli* and *Salmonella typhimurium*
Mutat. Res., 151, 201–207 (1985)

Perry, D.F., Carter, D.E., Sipes, I.G.
Distribution and excretion of ^{14}C-parachloroaniline in the rat
Toxicologist, 1, 14 (1981)

Pienta, R.J., Poiley, J.A., Lebherz III, W.B.
Morphological transformation of early passage golden Syrian hamster embryo cells derived from cryopreserved primary cultures as a reliable in vitro bioassay for identifying diverse carcinogens
Int. J. Cancer, 19, 642–655 (1977)

Pienta, R.J., Kawalek, J.C.
Transformation of hamster embryo cells by aromatic amines
Natl. Cancer Inst. Monogr., 58, 243–251 (1981)

Ping Pan, H., Fouts, J.R., Devereux, T.R.
Hepatic microsomal N-hydroxylation of p-chloroaniline and p-chloro-N-methylaniline in red-winged blackbird compared with rat
Xenobiotica, 9 (7), 441–446 (1979)

Prasad, I.
Mutagenic effects of the herbicide 3′,4′-dichloropropionanilide and its degradation products
Canad. J. Microbiol., 16, 369–372 (1970)

Rankin, G.O., Yang, D.J., Cressey-Veneziano, K., Casto, S., Wang, R.T., Brown, P.I.
In vivo and in vitro nephrotoxicity of aniline and its monochlorophenyl derivatives in the Fischer 344 rat
Toxicology, 38, 269–283 (1986)

Rashid, K.A., Arjmand, M., Sandermann, H., Mumma, R.O.
Mutagenicity of chloroaniline/lignin metabolites in the Salmonella/microsome assay
J. Environ. Sci. Health, B22 (6), 721–729 (1987)

Rosenkranz, H.S., Poirier, L.A.
Evaluation of the mutagenicity of DNA-modifying activity of carcinogens and noncarcinogens in microbial systems
J. Natl. Cancer Inst., 62 (4), 873–891 (1979)

Sakagami, Y., Yamazaki, H., Ogasawara, N., Yokoyama, H., Ose, Y., Sato, T.
The evaluation of genotoxic activities of disinfectants and their metabolites by umu test
Mutat. Res., 209, 155–160 (1988)

Sakagami, Y., Yamazaki, H., Ogasawara, N., Yokoyama, H., Ose, Y., Sato, T.

Evaluation of genotoxic activities of disinfectants and their metabolites by the umu test
Mutat. Res., 216, 373 (1989)

Scholes, E.W., Basketter, D.A., Sarll, A.E., Kimber, I., Evans, C.D., Miller, K., Robbins, M.C., Harrison, P.T.C., Waite, S.J.
The local lymph node assay: results of a final inter-laboratory validation under field conditions
J. Appl. Toxicol., 12 (3), 217–222 (1992)

Scott, A.I., Eccleston, E.
Investigations of the general toxic and haematological effects of para-chloroaniline in several species
Proc. Eur. Soc. Drug Tox., 8, 195–204 (1967)

Scotti, P., Tomasini, M.
Su di un caso di grave intossicazione acuta da paracloroanilina con intensa metemoglobinemia e con transitorie alterazioni elettrocardiografiche (Italian with English summary)
Med. Lav., 51 (1), 662–666 (1966)

Seuferer, S.L., Braymer, H.D., Dunn, J.J.
Metabolism of diflubenzuron by soil microorganisms and mutagenicity of the metabolites
Pest. Biochem. Physiol., 10, 174–180 (1979)

Simmon, V.F.
In vitro mutagenicity assay of chemical carcinogens and related compounds with *Salmonella typhimurium*
J. Natl. Cancer Inst., 62 (4), 893–899 (1979 a)

Simmon, V.F.
In vitro assays for recombinogenic activity of chemical carcinogens and related compounds with *Saccharomyces cerevisiae*

Butynediol

1. Summary and assessment

Based on the available acute toxicity studies, butynediol is toxic on oral administration (LD_{50} rat oral 100 to 176 mg/kg body weight) and moderately toxic following dermal absorption of aqueous solutions (LD_{50} rat dermal 659 mg/kg body weight). The main clinical signs of toxicity are apathy, disturbances of balance, convulsions and diarrhoea, the predominant effects at post-mortem being congestion of the internal organs, pulmonary oedema and haemorrhage and fatty infiltration of the liver. The results of exposure in the inhalation hazard test are dependent on the temperature at which the atmosphere is produced and the duration of exposure. Thus, all 12 rats exposed to an atmosphere enriched or saturated with butynediol at 20 °C for 8 h survived, while exposure to an atmosphere produced at 70 °C was tolerated for 2 h but was fatal to all 6 rats after exposure for 8 h. In mice exposed to the volatile vapours of butynediol at 35 to 40 °C for 2 h, signs of irritation of the mucous membranes have been seen, as well as motor excitation with subsequent depression and deaths. Post-mortems and examination by light microscope revealed congestion of the internal organs and the brain, haemorrhages and bronchitis in the lungs and tubular degeneration in the kidneys.

No treatment-related effects have been observed after administration of 5, 10 or 20 mg butynediol/kg body weight to rats by gavage for 5 days, except for an increase in the cholesterol level in males at the top dose level only,which was not dose dependent. No pathological effects have been shown in tests of nervous system function. The no effect levels following oral administration to rats are 10 mg/kg body weight after 2 weeks and 1 mg/kg body weight after 4 weeks, with congestion of the internal organs, pulmonary oedema and diffuse parenchymal necrosis and fatty degeneration of the liver at higher doses. Haematological effects in the form of anaemia, and tubular degeneration and interstitial cell infiltration in the kidneys have also been described. Similar effects (with the exception of liver damage) have also been observed in mice exposed for 1 month to butynediol vapour produced at 30 °C (ca. 0.09 to 0.12 mg/l air) for 2 h/day.

Numerous skin and eye irritation studies have been carried out in rabbits, with butynediol in various preparations and at various dilutions, giving differing results. In general, the substance is irritating to the skin and, at most, very slightly irritating to the eye.

Butynediol has not shown any skin sensitizing effects in three valid studies in guinea-pigs.

No characteristic signs of toxicity occurred in rats on long-term oral administration of 0.04 and 0.2 mg/kg body weight/day for 6 months. At 2 mg/kg body weight/day, the main effects were inhibition of conditioned reflexes, patchy congestion of organs and toxic effects on the liver. The no effect level in this study is 0.2 mg/kg body weight/day. The inhalation of 0.01 mg butynediol/l air as an aerosol by rats for 6 months was said to have caused marginal effects on the respiratory tract that were reversible after 1 month.

The studies involving repeated oral administration, in which the no effect level is clearly reduced with increasing study duration, indicate a cumulative effect of butynediol.

Butynediol is not mutagenic in the Salmonella/microsome test and does not cause chromosome aberrations in Chinese hamster V79 cells.

Butynediol has shown no tumour initiating effect in an early dermal initiation/promotion study in mice, following application of butynediol once a week for 10 weeks and croton oil once a week for 18 weeks.

In a preliminary study prior to an embryotoxicity/teratogenicity study, the administration of 20, 40 and 60 mg butynediol/kg body weight/day to pregnant Wistar rats by gavage from day 6 to day 15 p.c. led to dose-dependent maternal toxicity. Body weight gain and feed intake were significantly reduced at the top dose level at the beginning of the study and a dose-dependent disturbance of liver function was observed in all three dose groups. Based on these results, doses of 10, 40 and 80 mg butynediol/kg body weight/day were chosen for the main study. In this study, signs of maternal toxicity seen in the high dose group included reduced feed intake and body weight loss, apathy, poor general condition and vaginal haemorrhages. One high-dose animal died, and macroscopic effects seen at post-mortem in this rat included a stained liver and marginal pulmonary emphysema. An increase in the number of foetuses per litter with an accessory 14th rib, seen at the high dose level, has been evaluated as a marginal sign of developmental toxicity in the embryos, possibly a manifestation of non-specific stress on the dams, but not a teratogenic effect. The no observed adverse effect level was given as 40 mg/kg body weight for the dams and the foetuses.

Butynediol can lead to a reduction in body temperature in rats, possibly caused by peripheral vasodilation.

In man, dermatitis caused by butynediol and two cases of contact allergy have been described.

Acute, 5-day and 4-week inhalation studies in rats, including tests for neurotoxicity, will be carried out on behalf of BG Chemie.

2. Name of substance

2.1	Usual name	Butynediol
2.2	IUPAC-name	2,4-Butyne-1,4-diol
2.3	CAS-No.	110–65–6
2.4	EINECS-No.	203–788–6

3. Synonyms, common and trade names

Agrisynth B3D
Bis(hydroxymethyl)acetylene
Butindiol
Butyne-2-diol-1,4
But-2-yne-1,4-diol
1,4-Butynediol
2-Butynediol
2-Butynediol-1,4

		1,4-Dihydroxy-butyne-2 1,4-Dihydroxy-2-butyne Golpanol Korantin BH

4. Structural and molecular formulae

4.1	Structural formula	HO–CH_2–C≡C–CH_2–OH
4.2	Molecular formula	$C_4H_6O_2$

5. Physical and chemical properties

5.1	Molecular mass, g/mol	86.09
5.2	Melting point, °C	57.5 (Sax, 1979) 54–56 (BASF, 1993)
5.3	Boiling point, °C	238 (Fliege et al., 1975; BASF, 1993) 125–127 (at 3 hPa) (BASF, 1993)
5.4	Vapour pressure, hPa	2.3 (at 120 °C) (Fliege et al., 1975) 1.33 (at 102 °C) (BASF, 1993)
5.5	Density, g/cm^3	1.17 (at 20 °C) (Falbe and Regitz, 1989)
5.6	Solubility in water	Dissolves well: 374 g/100 g water (at 25 °C) (Fliege et al., 1975)
5.7	Solubility in organic solvents	Dissolves readily in alcohols and acetone; dissolves with difficulty in ethers and hydrocarbons; practically insoluble in benzene (Fliege et al., 1975)
5.8	Solubility in fat	No information available

5.9 pH value	ca. 3 (at 100 g/l water) (BASF, 1987 a)
5.10 Conversion factor	1 ml/m^3 (ppm) ≙ 3.51 mg/m^3 1 mg/ ≙ 0.28 ml/m^3 (ppm) (at 1013 hPa and 25 °C)

6. Uses

As a glazing agent in electroplating, corrosion inhibitor, stabilizer for halogenated hydrocarbons; intermediate in the manufacture of 2-butene-1,4-diol, 1,4-butanediol, insecticides and herbicides (Gräfje et al., 1985).

Intermediate in the chemical industry, for example in the manufacture of pharmaceuticals, plant protection products and pesticides, textile processing aids, corrosion inhibitors, plasticizers, plastic resins and stabilizers; additive in nickel plating baths (BASF, 1980).

7. Experimental results

7.1 Toxicokinetics and metabolism

In the rat, butynediol is oxidized by alcohol dehydrogenase to degradation products which have not yet been further defined. The Michaelis constant (K_m) for this reaction under in vitro conditions is 8.2×10^{-4} M (for comparison, that for ethanol is 7.9×10^{-4} M; Taberner and Pearce, 1974).

7.2 Acute and subacute toxicity

Acute toxicity

Data on the acute toxicity of butynediol are summarized in Tables 1, 2 and 3. The LD_{50} values in mice (see Table 1) following oral administration are 100 and 105 mg/kg body weight and after intraperitoneal injection are between 15 and 100 mg/kg body weight, while the LD_{50} on subcutaneous injection is 63 mg/kg body weight. In the rat (see Table 2) oral LD_{50} values of between 100 and 176 mg/kg body weight have been found, with males being more susceptible, while after intraperitoneal injection LD_{50} values of between 52 and 55 mg/kg body weight have been reported and after dermal application values between 424 and 1240 mg/kg body weight were found. The oral LD_{50} in the rabbit (see Table 3) was 150 mg/kg body weight, while that in the guinea pig was 130 mg/kg body weight. Signs of toxicity on acute exposure included sedation, analgesia, disturbances of balance, lying on the side, tonic-clonic convulsions, ruffled fur, accelerated breathing, bradycardia, apathy, salivation and, in rats, coughing (according to the authors; Taberner and Pearce, 1974), hyperaemia of the visible skin, reductions in body temperature and diarrhoea. Post-mortems revealed pulmonary haemorrhages, petechiae of the mucous membrane of the gastro-intestinal tract, oedema and congestion of the internal organs, and signs of fatty degeneration of the liver. The substance should thus be evaluated as being toxic on oral administration and of moderate toxicity on dermal application.

In a preliminary study, a single oral dose of 50 mg/kg body weight was lethal to cats, while the same dose given either once or twice was survived by both exposed rabbits, although accelerated breathing and diarrhoea were seen. A single dose of 100 mg/kg body weight caused death in rabbits. Macroscopic examination revealed isolated bleeding in the gastro-intestinal area, fatty degeneration of the liver and pulmonary oedema (BASF, 1959 a, b, 1987 b, 1993).

In another study, groups of 5 male and 5 female Wistar rats (322 43 g and 209 21 g) were given a single dose of butynediol (ca. 99% pure) at doses of 100, 150, 180, 200 and 250 mg/kg body weight as a 10% aqueous solution by gavage. The observation period was 14 days. The calculated LD_{50} values were 132 (89 to 158) mg/kg in males and 176 (118 to 270) mg/kg body weight in females. The rats died within 48 h of dosing. No clinical signs of toxicity were described. Macroscopic effects in the rats that died included diarrhoea and congestion of the internal organs, while the main histological effects were liver and kidney damage. These were also observed in the rats killed at the end of the observation period (Jedrychowski et al., 1992 a).

The acute dermal toxicity was investigated in male and female Wistar rats of the Hoe:WISKf(SPF71) strain after a single 24-hour application and with a 14-day observation period in accordance with OECD guideline no. 402. The purity of the substance tested in 2 separate experiments was 98.9% or 99.5% and the doses ranged from 50 to 2000 mg butynediol/kg body weight (made into a paste with 0.9% aqueous saline). Signs of toxicity observed included impaired breathing and coordination, partly-closed eyes, blood-stained encrustation of the muzzle and margins of the eyelids and loss of righting and toe-pinch reflexes. Body weight gain was retarded in the first week of the study. The fatally poisoned rats died within 3 days. At autopsy, accumulation of blood in the colon and bladder, marks on the liver lobes and staining of the liver and spleen were seen. The LD_{50} value in male rats was 424 mg/kg body weight while that in female rats was 983 mg/kg body weight. An LD_{50} of 659 mg/kg body weight was calculated for both sexes together. Male rats were thus more susceptible than females (Hoechst, 1990).

The acute dermal toxicity was tested in a further study in which butynediol (ca. 99% pure) was applied under occlusive cover to the clipped dorsal skin of Wistar rats (213 ± 17 g) either in solid form (moistened with water; 11 rats) or as a 40% aqueous solution (16 rats) as a single dose of 5000 mg/kg body weight. The duration of exposure was 24 h and the observation period was 14 days. No deaths occurred following the application of butynediol in solid form (as a paste in water) while after application of the 40% aqueous solution, 8/16 rats died within 48 h. Post-mortems on these rats revealed liver and kidney damage. At the end of the 14-day observation period, liver damage was evident both in the rats treated with the 40% solution and in those treated with the paste (Jedrychowski et al., 1992 a).

No mortality was observed in 6 rats exposed to atmospheres enriched or saturated with butynediol at 70 °C for 2 h in an inhalation hazard test. Exposure for 8 h was lethal to all 6 exposed rats after $4\,^1/_4$ to 24 h (BASF, 1959 a).

In an inhalation hazard test with atmospheres enriched or saturated with butynediol at 20 °C, all 12 rats survived an 8-hour exposure without showing signs of toxicity (BASF, 1973).

A group of 20 mice was exposed for 2 h to the volatile vapours of butynediol at 35 to 40 °C ("dense white vapours") at a concentration of 0.15 to 0.28 mg/l air. Signs of irritation of the mucous membranes could be observed (closed eyes, wiping of muzzles) as well as motor excitation with subsequent depression. Deaths occurred both during and after exposure (no further details). The surviving animals recovered only slowly. Post-mortems on the mice that died revealed congestion of the internal organs and the brain. The lungs showed uneven congestion and small haemorrhages as well as marked catarrhal-desquamative bronchitis. After the 14-day observation period, histological effects included congestion and oedema of the internal organs, an increase in polynuclear leukocytes in the lungs and spleen and moderate albuminous degeneration of the convoluted tubules of the kidney (Stasenkova and Kochetkova, 1965).

The lowest published lethal concentration (LC_{Lo}) for a 2-hour inhalation exposure to butynediol was 150 mg/m^3 (equivalent to 42 ppm) in mice and rats (no further details; Izmerov, 1982).

Subacute toxicity

In a 5-day study, groups of 5 male and 5 female Wistar rats were given 5, 10 or 20 mg butynediol/kg body weight/day by gavage (98.9% pure, dissolved in double-distilled water). Neurological function was tested in all animals before the study began, 24 h after the first dose and 24 h after the fifth dose. In addition to thorough observation of the animals, the following sensory and motor tests were carried out: sensitivity to touch, coordination of movement (righting response), eye blink response, pupil response, vision (visual placing response), hearing (startle response), olfactory ability, forelimb and hindlimb grip strength, toe-pinch, tail-pinch and pain perception (hot plate test). Biochemical and haematological parameters were studied in accordance with OECD guideline no. 407; in addition cholinesterase activity was determined in the serum and erythrocytes. The weights of the liver, kidneys, adrenal glands and testes were determined. Analysis of the concentrations of the dosing solutions showed that the amounts administered were 98.5% (equivalent to 19.7 mg/kg) at the top dose level, 87% (equivalent to 8.7 mg/kg) in the intermediate group and 80% (equivalent to 4 mg/kg) in the low dose group. Apart from a significant increase in cholesterol in the males at the top dose level only, no treatment-related effects were observed, and the no effect level in this 5-day study was thus 10 mg/kg body weight (BASF, 1992 a).

In order to study the subacute toxicity of butynediol (97 to 99% pure), groups of 10 male and 10 female Sprague-Dawley rats (200 to 300 g) were given 1, 10 or 100 mg/kg body weight as an aqueous solution by gavage on 14 consecutive days prior to being killed on day 15. After 8 doses, one male in the top dose group and one female in the low dose group died, although these deaths were judged not to be treatment-related (probably gavaging errors). Body weight gain was significantly reduced ($p \leq 0.05$) compared with the controls in the males but not the females in the top dose group (weight gain 51 ± 13 g in high dose males, 86 ± 16 g in controls). Some rats at the top dose level had bloody nasal secretions, piloerection and diarrhoea. A significant ($p \leq 0.05$) increase in serum cholesterol occurred in both sexes in the top dose group (males 77.1 mg/100 ml, controls 60.0 mg/100 ml; females 112.1 mg/ml, controls 65.3 mg/ml), with an increase in serum calcium found in the females in this group (11.2 mg/ml, controls 10.6 mg/ml) as well as a fall in aspartate aminotransferase (glutamate-oxaloacetate transaminase) activity (106.5 mU/ml, controls 144.8 mU/ml) and in glucose level (154.6 mg/100 ml, controls, 177.5 mg/100 ml). Females in the high dose group also showed a slight increase in aminopyrine demethylase activity (14.8 nmol HCHO/hour/mg protein, controls 11.0 nmol HCHO/hour/mg protein). A fall in the erythrocyte count occurred in females ($6.8 \times 10^6/\mu l$, controls $7.5 \times 10^6/\mu l$) along with a fall in the haemoglobin content (13.1 g/100 ml, controls 14.4 g/100 ml). No significant effects on the red blood cell count were found in the males. At post-mortem on day 15, the relative liver weights were significantly increased ($p \leq 0.05$) in the rats in the top dose group (males 5.5 0.5%, controls 3.6 0.3%; females 5.3 ± 0.5%, controls 3.5 ± 0.4%). Histological examination of 28 organs revealed no treatment-related effects. The no effect level following oral administration for 14 days was thus 10 mg/kg body weight (Komsta et al., 1989).

In a 4-week study, groups of 8 male and 8 female Wistar rats (average starting weights 188 ± 20 g and 145 ± 15 g, respectively) were given doses of 0, 1, 10 and 50 mg butynediol/kg body weight as an aqueous solution once daily by gavage. In the top dose group, 3 males and 3 females died after 7 to 28 days, most after 26 to 28 days. Body weight gain was significantly ($p < 0.05$) reduced in the males, while that of the females was unaffected. Haematological effects at the end of treatment in the females in the top dose group included significant reductions in the erythrocyte count ($p \leq 0.01$), the haemoglobin concentration ($p \leq 0.05$) and the haematocrit value ($p \leq 0.01$) and there was a significant increase in the reticulocyte count ($p \leq 0.01$) in both sexes. The leukocyte count was significantly increased ($p \leq 0.01$) in both sexes at the top dose level, with neutrophils ($p \leq 0.05$ and 0.01) and lymphocytes (p 0.05) affected. In clinical chemistry tests at the end of the study, the activity of sorbitol dehydrogenase was significantly increased ($p \leq 0.01$) in both sexes at the top dose level. The total protein content in the serum was significantly increased in females in the top dose group ($p \leq 0.05$) and the glucose level was significantly increased in males in the top dose group ($p \leq 0.05$). At the end of the study, there were increases in the absolute and relative liver weights ($p \leq 0.01$) and in the absolute and relative kidney weights ($p \leq 0.05$ to $p \leq 0.01$) in both sexes at the top dose level. The relative liver weight was also increased in females at the intermediate dose level ($p \leq 0.05$). Histological examination of rats at the high and intermediate dose levels revealed congestion of the internal organs, pulmonary oedema, marked tubular degeneration and interstitial mononuclear cell infiltration in the kidneys, and severe diffuse parenchymal necroses in the liver (mainly centrilobular) with fatty degeneration in the rest of the parenchyma, effects which were absent in the low dose group. The no effect level in this study was thus 1 mg/kg body weight (Jedrychowski et al., 1992 b).

A group of 20 mice (18 to 20 g) inhaled butynediol vapour, which was produced by heating the substance at 30 °C, for 2 h daily, 6 times a week for one month. The concentration was between 0.09 and 0.12 mg/l air. Clear signs of irritation of the eyes and respiratory organs were observed. Body weight was retarded by 15 to 20% compared with the control group (10 mice) 14 days after the beginning of the study, an effect that continued during the last 2 weeks of the study. At the end of the second week, the neuromuscular stimulation threshold had fallen to 5 mA compared to 8 to 9 mA in the control group. On day 18 of the study, 2 animals died during exposure. Postmortems on mice killed immediately after the end of the last exposure revealed severe hyperaemia of the internal organs and the brain. Microscopic effects included congestion and mild oedema of the internal organs, moderately pronounced pulmonary emphysema, an increase in polynuclear leukocytes in the spleen and albuminous degeneration of the convoluted tubules in the kidney (Stasenkova and Kochetkova, 1965).

7.3 Skin and mucous membrane effects

Skin irritation studies

The occlusive application of a 30% aqueous solution of butynediol to the skin or to the ear of the rabbit was not irritating (duration of application 1, 5 or 15 min or 20 h; BASF, 1959 a).

When pure butynediol was applied for 2 h to a 25 cm^2 area of shaved skin in rabbits (number unspecified), pain-sensitive severe reddening and hardening of the skin occurred after exposure. The symptoms cleared up after 1 to 2 days and after 3 to 5 days, scabs formed, which sloughed off after 5 to 7 days. Flaking of the skin was also observed. When 0.5 ml of a 30% aqueous solution of butynediol was applied to a further animal, marked reddening, which was completely reversible after 2 to 3 days, also occurred (Stasenkova and Kochetkova, 1965).

An 80% aqueous preparation of butynediol was applied to the skin of rabbits for 1, 5 or 15 min or for 20 h. The 20-hour exposure led to slight reddening and severe oedema after 24 h. After 8 days, severe necrosis was observed with leather-like changes and reddened skin. The substance was thus corrosive in this study (BASF, 1973).

On semi-occlusive application of an 80% aqueous formulation (w/w) of neat butynediol to the dorsal skin of rabbits (Vienna white; 2 male, 1 female) for 4 h in accordance with OECD guideline no. 404, redness and oedema occurred after 24 h, haemorrhages and moderate to severe overlapping oedema after 48 to 72 h, and flaking of the skin after 8 days. The substance thus had an irritant effect (BASF, 1986 a).

In a further study of the primary skin irritancy of ca. 99% pure butynediol, 4 rabbits (Vienna white; 3.8 to 4.2 kg) had either 0.3 g of the solid substance moistened with water or 40 or 20% aqueous solutions, applied occlusively to the clipped intact or scarified flank skin (2 cm x 2 cm). The exposure time was 24 h. The findings were recorded after 1, 24, 48 and 72 h and evaluated according to the Draize scheme. The solid butynediol caused no irritation of the intact skin, although when applied to the scarified skin, 1 of 4 rabbits showed slight reddening. With the 40% solution, the skin of one of 4 rabbits also showed slight redness after treatment, while the 20% solution did not cause skin irritation in any of the rabbits. Based on these results, the authors evaluated butynediol as not irritating to the skin (Jedrychowski et al., 1992 a).

Repeated application of a 30% solution of butynediol to the skin of the rabbit for 10 days caused severe reddening, hardening, scab formation and sloughing of the epidermis (no further details; Stasenkova and Kochetkova, 1965).

Eye irritation studies

No irritation was observed following instillation of a 30% aqueous solution of butynediol into the rabbit eye (no further details; BASF, 1959 a, 1987 b).

Lacrimation, pronounced hyperaemia of the sclera and partly closed eyelids were observed after instillation of one drop of a 30% aqueous solution of butynediol into the conjunctival sac of a rabbit. After the second instillation, oedema of the mucous membrane of the eye was observed (no data on time of second instillation) and on the following day, purulent conjunctivitis developed. The effects described were completely reversible after 6 to 8 days (Stasenkova and Kochetkova, 1965).

In an inadequately documented study, it was reported that butynediol was not irritating to the rabbit eye at concentrations below 1 mg/l water, although it could be irritant at higher concentrations (no further details; Knyshova, 1968).

When 50 mm^3 or 50 mg butynediol (no further details) were instilled once into the conjunctival sac of the rabbit (no data on the number of rabbits exposed), slight redness, severe oedema and a viscous surface layer were seen after 1 and 24 h. The effects were completely reversible after 8 days, and butynediol was therefore not irritating in this study (BASF, 1973).

In a further study, which was carried out in accordance with OECD guideline no. 405, the introduction of the solid substance into the rabbit eye (Vienna white; 2 males, 1 female) caused only slight redness of the conjunctiva, which was completely reversible after 72 h. Butynediol did not therefore have an irritant effect (BASF, 1986 b).

In another study, 100 mg ca. 99% pure butynediol was introduced into the conjunctival sac of one eye in each of 4 New Zealand white rabbits (no data provided on body weight). The findings were recorded and evaluated according to the Draize scheme after 1, 24, 48 and 72 h and after up to 7 days. After one hour, butynediol caused lacrimation and slight closing of the eyes in all of the rabbits. Minimal conjunctival erythema, which was reversible later in the study, was evident after 24 and 48 h (Jedrychowski et al., 1992 a). Butynediol was thus not irritating to the eye in this study.

7.4 Sensitization

In a study of the sensitizing properties of butynediol, which was not carried out according to current guidelines, 7 white rabbits (2 of which were controls) were treated with neat butynediol, 0.5 g of which was applied to a 4 cm x 4 cm area of clipped skin every day for 10 days, with the untreated skin on the other side of the body serving as a control. The 10 applications were repeated 3 times at intervals of 15 to 20 days, with a new skin section on the same side of the body being treated each time. In the first series of treatments, the skin irritation became more intense after 5 to 7 applications, with an increase in hyperaemia and haemorrhages seen, as well as tears in the skin. After 8 to 10 applications, marked xeroderma, skin thickening and profuse flaking were observed, with the skin appearing normal again after 15 to 20 days. In the second series of treatments, the skin effects occurred on day 2 to day 3 and were more marked. The skin reaction in the third series of treatments was somewhat weaker again. Histopathological examination of the treated rabbit skin after the first series of 10 applications showed congestion of the skin with haemorrhages and round cell infiltration around vessels and hair follicles. On examination after the end of the second and third series of treatments, focal thinning of the epidermis was observed with partial shedding of the skin and diffuse round cell infiltration. The authors concluded from the results that butynediol had both a direct skin irritant effect and a skin sensitizing effect (Stasenkova and Kochetkova, 1965). The study was not carried out in accordance with a validated protocol. The more pronounced skin reaction in the second series of treatments could possibly be explained by varying sensitivity of the different areas of skin treated and is not an unequivocal indication that the substance has sensitizing potential.

Butynediol (99.2% pure) was tested for skin sensitization potential in female guinea-pigs (281 to 379 g at the start of the study) as a 25% solution in physiological saline in a maximization test in accordance with OECD guideline no. 406. Very slight erythema developed after 24 h during the induction phase in both the control (5/20) and the test animals (14/18, 2 animals died during the study). No reactions occurred in the controls after challenge treatment, while in the treated animals a slight erythematous reaction was observed in 1/18 guinea pigs after 24 h, an effect which was viewed as coincidental by the authors. Butynediol had no skin sensitizing properties in this study (RCC, 1990).

Butynediol also had no skin sensitizing effect in another study, a maximization test carried out according to Magnusson and Kligman in male white Hartley guinea pigs (403 ± 48 g; 22 treated animals, 8 controls). Based on preliminary studies, intradermal induction was carried out with a 2% solution, dermal induction with a 20% solution and challenge with 5 and 20% solutions. Allergic contact dermatitis did not occur in any of the test animals (no further details; Jedrychowski et al., 1992 a).

In another Magnusson/Kligman study, the ability of butynediol (solid, > 99% pure) to induce skin sensitization was studied in 20 male albino guinea pigs compared with 10 control animals in accordance with OECD guideline no. 406. Intracutaneous induction was carried out with 0.5% butynediol (dissolved in liquid paraffin), epidermal challenge treatment with 25% butynediol in water. A positive reaction was seen in 5 of 20 animals, and a weak sensitizing potential was attributed to the substance (Hüls, 1985). In accordance with "EC labelling guidelines", the substance was evaluated as not sensitizing (BASF, 1993).

7.5 Subchronic and chronic toxicity

In a 6-month oral study, the administration of butynediol to male rats (6/group) at doses of 0, 0.04, 0.2 and 2 mg/kg body weight/day did not lead to any effects on general behaviour, body weight or blood parameters (haemoglobin level, numbers of erythrocytes, leukocytes and thrombocytes and coagulation time). At 2 mg/kg body weight/day, delayed development of conditioned reflexes occurred with an increase in the latent period by 40%. In addition, reduced cholinesterase and SH-enzyme activities and increased transferase (transaminase) activities were seen, together with an altered serum protein profile. In the brain, the high dose led to a reduction in the number of Nissl's bodies and an increase in neuroglia as well as reduced SH-enzyme activity. In the liver, fatty degeneration, sclerotic areas and reduced glycogen levels were found, while in other organs, localized hyperaemia occurred. The 0.2 and 0.04 mg/kg body weight/day doses were without effect (no further details; Knyshova, 1968).

In a 6-months inhalation study, 20 rats (180 to 200 g) were exposed to the aerosol of a 5% aqueous solution of butynediol for 4 h daily, 6 times a week. The concentration was 0.008 to 0.01 mg/l air (droplet size ≤ 2 µm). The control group consisted of 20 rats. General condition and body weight gain were unaffected. After 4 months, the neuromuscular stimulation threshold had fallen to 6.0 ± 0.2 mA compared with 8.9 ± 0.1 mA in the controls. The difference was also detectable in months 5 to 6 of the study ($p < 0.02$). Towards the end of month 6, the haemoglobin level was slightly reduced at 78.3 ± 1.2% compared with 85.2 ± 1.3% in the controls, while the numbers of erythrocytes and leukocytes were within the normal physiological range. At the end of the 6th month of exposure, the arterial blood pressure was on average 70.0 ± 3.5 mmHg compared with 98.2 ± 1.5 mmHg in the controls ($p < 0.01$). The amount of urine produced daily was almost the same in the treated and control rats at 6 ± 1.2 ml, as was the protein content of the urine (57.4 ± 4.4 mg on average). In contrast, the daily excretion of hippuric acid was increased towards the end of month 6 from 128 ± 3.8 mg in the controls to 158 ± 7.2 mg ($p < 0.05$). At the end of the study, half of the rats were killed. There were no conspicuous macroscopic effects, although microscopic examination revealed minimal catarrhal bronchitis and slight hypertrophy of the musculature of the bronchial walls and their arteries. Hyperplasia of the lymphatic tissue was present around the large bronchi, and the alveolar walls were slightly thickened due to proliferation of lymphocytic and histiocytic cells and polynuclear lymphocytes. There were no particular histologically abnormal findings in the other organs examined. The remaining rats were killed one month after the end of the study. On post-mortem and histological examination, treatment-related effects were no longer evident, and the changes reported immediately after the end of the study were thus reversible. The concentration of 0.008 to 0.01 mg/l air was therefore described as the "threshold concentration" by the authors. Based on these studies, they recommended a provisional workplace limit value of 0.001 mg/l, equivalent to 1 mg/m^3 air (Stasenkova and Kochetkova, 1965).

7.6 Genotoxicity

7.6.1 In vitro

Butynediol (> 99% pure, dissolved in DMSO) was tested for mutagenic activity in the *Salmonella typhimurium* strains TA 1535, TA 1537, TA 1538, TA 98 and TA 100 at concentrations of from 20 to 5000 mg/plate with and without metabolic activation (S9-mix from Aroclor 1254-induced rat liver). Butynediol showed neither cytotoxic nor mutagenic effects in this study (BASF, 1981).

An in vitro test for structural chromosome aberrations was carried out with 99.5% pure butynediol in Chinese hamster V79 cells. The concentrations tested were 860, 300 and 50 µg/ml without metabolic activation and 300, 100 and 10 µg/ml with metabolic activation (S9-mix from Aroclor 1254-induced rat liver). Chromosomes were prepared after 7, 18 and 28 h, and 100 to 200 metaphases were evaluated per culture. Cell growth was inhibited at concentrations of 860 µg/ml (without activation) and 300 µg/ml (with activation). Chromosome aberrations did not occur at any of the tested concentrations. Thus, in this study, butynediol had no genotoxic activity (CCR, 1989/1991).

7.6.2 In vivo

No information available.

7.7 Carcinogenicity

White mice (groups of 10, strain S, 7 to 9 weeks old) were treated with a total dose of 4 or 40 mg butynediol (dissolved in acetone, volume applied 0.3 ml), which was applied to their clipped dorsal skin once weekly for 10 weeks. In addition, the mice had croton oil applied to the skin once a week for 18 weeks (0.5% in acetone, volume applied 0.3 ml), beginning 3 days after the first substance application. 3 control groups, each consisting of 20 mice, were treated only with croton oil for 18 weeks. The treated area was inspected weekly for tumour growth, with tumours of 1 mm and above included. One week after the end of treatment (week 19), 57/60 of the control mice were still alive. In 3/57 mice, a total of 7 tumours had occurred; these were shown to be benign papillomas on histopathological examination. With additional butynediol application, 19/20 of the mice had survived (no details of the cause of death of the one mouse that died in the low dose group), and one mouse in the low dose group had developed a benign papilloma. The mice in one of the 3 control groups were observed for the remainder of their lives after the end of the 18-week croton oil treatment, although no further tumours developed and no malignant changes to the papillomas occurred. No increase in the incidence of tumours was found in the lung, which was examined histopathologically in the butynediol groups. Thus butynediol had no tumour-initiating effect in this initiation/promotion study with croton oil (no further details; Roe, 1957). It should be noted that the study duration, at 19 weeks, is too short to be able to make a definite statement on the initiating effect of butynediol.

7.8 Reproductive toxicity

In a preliminary study which was carried out in accordance with OECD guideline no. 414 prior to an embryotoxicity/teratogenicity study in order to find a dose causing maternal toxicity, groups of 5 pregnant Wistar rats were given 20, 40 and 60 mg butynediol/kg body weight/day (98.9% pure, dissolved in double-distilled water) by gavage from day 6 to day 15 p.c.. The following statistically significant signs of maternal toxicity were found in the high dose group on day 16 p.c.: reductions in feed intake and body weight loss from days 6 to 8 p.c., increases in total protein and albumin, reductions in thromboplastin time, serum cholinesterase activity and triglycerides, and increases in the absolute and relative kidney and liver weights. There was also a significant increase in the albumin content of the blood, significant decreases in thromboplastin time and serum cholinesterase activity and significant increases in absolute and relative liver weights in the intermediate dose group. In the low dose group there was a significant increase in albumin, and dose-dependent maternal toxicity was thus observed. There were no differences in the reproductive parameters studied between the treated and control groups. Based on these results, doses of 10, 40 and 80 mg butynediol/kg body weight/day were chosen for the subsequent embryotoxicity/teratogenicity study (BASF, 1992 b).

Groups of 18 to 22 pregnant Wistar rats were given 10, 40 and 80 mg butynediol/kg body weight (98.9% pure, volume 10 ml/kg body weight) as an aqueous solution by gavage from day 6 to day 15 p.c., while the control group (22 pregnant females) was given the vehicle only. The study was carried out in accordance with OECD guideline no. 414. On day 20 p.c., the rats were examined macroscopically, and the foetuses were removed and weighed, their sex was determined and they were thoroughly examined. Signs of maternal toxicity seen in the high dose group included a statistically significant reduction in feed intake in the first half of the study period from day 6 to day 8 p.c. (ca. 21% less than in the control rats), a statistically significant loss of body weight between day 6 and day 8 p.c., and one death on day 8 p.c., with vaginal haemorrhages, apathy, poor general condition and piloerection seen prior to death. Macroscopic effects seen at post-mortem in the rat that died included a stained liver and marginal pulmonary emphysema. In another rat, piloerection was observed on day 8 and day 9 p.c.. There was an increase in the number of foetuses per litter with an accessory 14th rib, in the high dose group. This was evaluated by the authors as a marginal sign of developmental toxicity in the embryos, possibly a manifestation of non-specific stress on the dams, but not a teratogenic effect. In the intermediate and low dose groups there were neither signs of maternal toxicity nor embryo/foetotoxicity and thus the no observed adverse effect level (NOAEL) was 40 mg/kg body weight/day for the dams and the foetuses (BASF, 1994).

7.9 Effects on the immune system

No information available.

7.10 Neurotoxicity

See Sect. 7.2 "Subacute toxicity" (BASF, 1992 a).

7.11 Other effects

Butynediol led to marked hypothermia in rats. Intraperitoneal injection of 0.408 mmol butynediol/kg body weight (equivalent to 35 mg/kg body weight) caused a reduction in body temperature to ca. 35 °C 2½ h after administration, while after a dose of 1.634 mmol/kg body weight i.p. (equivalent to 141 mg/kg body weight), the body temperature decreased to 30.1 °C. The body temperature of the animals that were given the low dose returned to normal within 12 h. The rats given the high dose died within 2½ to 3 h of the injection (Taberner and Pearce, 1974).

To study the possible tumour-inhibiting effects of butynediol, 5 white mice were treated with 5 mg butynediol/kg body weight/day for 7 days by intraperitoneal injection (previously determined intraperitoneal LD_{50} 15 mg/kg), starting 24 h after the subcutaneous transplant of a Crocker sarcoma 180 into the axillary area. The administration of butynediol caused no inhibition of tumour growth compared with the control group (Carlson and Morgan, 1954).

The sleep- or narcosis-inducing effect of butynediol was studied in 10 male white rats (Sprague-Dawley or Holtzmann, weight 250 to 350 g), and compared with that of the saturated congener 1,4-butanediol. A dose of 5.5×10^{-3} mol of the substance/kg body weight was administered to the rats as an aqueous solution by intraperitoneal injection. The neuropharmacological effects were determined by observing behaviour in accordance with the Irwin screen and by studying other clinical effects. Signs of toxicity observed included a marked depression of motor activity, pronounced vasodilation of the extremities, profuse diarrhoea and death after 2 h (no further details; Sprince et al., 1966). The dose administered was equivalent to 473.5 mg/kg body weight and was thus almost 10-fold higher than the intraperitoneal LD_{50} in rats.

8. Experience in humans

Investigations of the health of workers who were in direct contact with butynediol revealed dermatitis, the intensity of which was dependent on the duration of contact and individual susceptibility (no further details; Stasenkova and Kochetkova, 1965).

A 54-year-old worker, who had been employed for years in the storeroom of a galvanizing department and who was involved in weighing out, packing and mixing various substances (e.g. solvents, paints, synthetic resins, hardeners, epoxy resins, isocyanates, acrylic resins, diaminodiphenylsulfone, as well as butynediol as an additive for nickel baths), developed an itchy dermatitis of the hands and lower arms, despite wearing rubber gloves. The patient was patch tested with a standard patch test tray, the clinic's own patch test series and an additional test series of 47 substances to which he had been exposed, at non-irritant concentrations. Only a 2% solution of butynediol caused a clear positive reaction after 48 h; this increased in severity after 72 h. The patient, who had to make 10% solutions of butynediol in water, reported exacerbation of his condition merely on opening butynediol containers. Of 50 workers who had had contact with increasing amounts of butynediol over 5 years, this was the only patient who developed contact allergy to butynediol (Malten, 1980).

A 41-year-old woman developed itchy eczema of the face, hands and forearms on contact with butynediol as a component of a cleaning agent. The symptoms appeared about 12 h after contact. Standard patch tests were negative. In contrast, a test with a 10% dilution of the cleaning agent was positive. When the various components of this product were tested, only butynediol (0.01% in water) produced a positive reaction; the purity of the butynediol tested was > 99.9%. Patch tests in 55 control subjects, in whom a 10-fold higher concentration of butynediol was tested, did not cause any positive reactions. The authors concluded that the patient had a contact allergic reaction to butynediol (Baadsgaard and Jørgensen, 1985).

The olfactory threshold in man has been given as 200 mg butynediol/l water (no further details; Knyshova, 1968).

9. Threshold limit values

No validated limit values are available.

Table 1. Acute toxicity of butynediol in the mouse

Species	Number of animals/dose	Sex	Route	Observation period	LD_0 (mg/kg body weight)	LD_{100} (mg/kg body weight)	LD_{50}/LC_{50} (mg/kg body weight)	Reference
Mouse	10	–	oral	14 days	50	200	100	Stasenkova and Kochetkova, 1965
Mouse	–	–	oral	–	–	–	104.75	Knyshova, 1968
Mouse	–	–	i.p.	–	–	–	15	Carlson and Morgan, 1954
Mouse	–	–	i.p.	7 days	–	–	84	BASF, 1959 a
Mouse	–	–	i.p.	7 days	–	–	ca. 100	BASF, 1973
Mouse	20	–	inhalation (2 h)	–	–	–	150–280 mg/m^3*	Stasenkova and Kochetkova, 1965
Mouse	–	–	inhalation (2 h)	–	–	–	150 mg/m^3 (LCLo)	Izmerov, 1982
Mouse	–	–	subcutaneous	7 days	–	–	63	BASF, 1959 a

– no data
i.p. intraperitoneal
LCLo lowest lethal concentration
* no precise details of the number of animals that died; butynediol heated to 35 to 40 °C

Table 2. Acute toxicity of butynediol in the rat

Species	Number of animals/dose	Sex	Route	Observation period	LD_0 (mg/kg body weight)	LD_{100} (mg/kg body weight)	LD_{50}/LC_{50} (mg/kg body weight)	Reference
Rat	–	–	oral	7 days	–	–	136	BASF, 1959 a
Rat	10	–	oral	14 days	50	200	100	Stasenkova and Kochetkova, 1965
Rat	–	–	oral	–	–	–	104.50	Knyshova, 1968
Rat	–	–	oral	7 days	–	–	0.13 ml/kg bw	BASF, 1987 b (≙135 mg/kg bw)
Rat	–	–	oral	7 days	–	–	ca. 100	BASF, 1973
Rat (Wistar)	10	♂, ♀	oral	14 days	–	–	♂ 132 ♀ 176	Jedrychowski et al., 1992 a
Rat	5	–	dermal (30% aqueous solution)	–	on exposure for 2.5 h	–	BASF, 1959a	
Rat	5	–	dermal (neat "technical" butynediol 30%)	–	on exposure for 1 h	on exposure for 4 h	–	BASF, 1959 a

Table 2. (continued)

Species	Number of animals/dose	Sex	Route	Observation period	LD_0 (mg/kg body weight)	LD_{100} (mg/kg body weight)	LD_{50}/LC_{50} (mg/kg body weight)	Reference
Rat (Wistar)	10	♂,♀	dermal (made into a paste with saline)	14 days	–	2000 (in ♀)	1240[1]	Hoechst, 1988
Rat	5–10	♂,♀	dermal	14 days	♂ 50 ♀ 400	♀ 2000[2]	659 (♂ 424, ♀ 983)	Hoechst, 1990
Rat (Wistar)	11	♀	dermal (in solid form, made into a paste with water)	14 days	5000	-	–	Jedrychowski et al., 1992 a
Rat (Wistar)	16	♀	dermal (as a 40% aqueous solution)	14 days	–	–	5000[3]	Jedrychowski et al., 1992 a
Rat	–	–	inhalation (2 h)	–	–	–	150 mg/m^3 (LCLo)	Izmerov, 1982

Table 2. (continued)

Species	Number of animals/dose	Sex	Route	Observation period	LD_0 (mg/kg body weight)	LD_{100} (mg/kg body weight)	LD_{50}/LC_{50} (mg/kg body weight)	Reference
Rat (Wistar)	6	–	i.p.	24 h	0.2 mmol/kg bw (≙ 17.2 mg/kg bw)	0.817 mmol/kg bw (≙ 70.38 mg/kg bw)	0.609–0.635 mmol/kg bw (≙ 52.4–54.7 mg/kg bw)	Taberner and Pearce, 1974

–	no data
bw	body weight
i.p.	intraperitoneal
1	in females: 2000 mg/kg bw only tested in males (= LD_{80} in males)
2	2000 mg/kg bw in males = LD_{80}
3	only dose tested; death of 8/16 rats within 48 h
LCLo	lowest lethal concentration

Table 3. Acute toxicity of butynediol in in guinea-pigs, rabbits and cats

Species	Number of animals/dose	Sex	Route	Observation period	LD_0 (mg/kg body weight)	LD_{100} (mg/kg body weight)	LD_{50}/LC_{50} (mg/kg body weight)	Reference
Guinea pig	–	–	oral	–	–	–	130	Knyshova, 1968
Rabbit	2	–	oral[1]	–	50	100	–	BASF, 1959 a
Rabbit	2	–	oral[2]	–	150	300	–	BASF, 1959 a
Rabbit	–	–	oral	–	–	–	150	Knyshova, 1968
Rabbit	3	–	dermal[3]	4 weeks	ca. 65 on exposure for 24 h[4]	–	–	BASF, 1959 b
Cat	2	–	oral[1]	–	–	50	–	BASF, 1959 a
Cat	2	–	oral[2]	–	–	150	–	BASF, 1959 a

– no data
1 pure butynediol
2 30% solution of butynediol
3 neat “technical” butynediol, 30%
4 also no deaths on 20-hour exposure of both ears

References

Baadsgaard, O., Jørgensen, J.
Contact dermatitis to butin-2-diol 1,4
Contact Dermatitis, 13, 34 (1985)

BASF AG, Gewerbehygienisch-Pharmakologisches Institut
Bericht über die toxikologische Prüfung von Butindiol rein
Unpublished report no. VII/256 (1959a)

BASF AG, Gewerbehygienisch-Pharmakologisches Institut
Bericht über die toxikologische Prüfung von Butindiol, ca. 30 %ig
Unpublished report no. VII/257 (1959b)

BASF AG, Gewerbehygiene und Toxikologie
Gewerbetoxikologische Vorprüfung von Golpanol BOZ fest
Unpublished report no. XXIII/128 (1973)

BASF AG
Data sheet Butin-2-diol-1,4 (1980)

BASF AG, Gewerbehygiene und Toxikologie
Bericht über die Prüfung von Butin-2-diol-1,4 im Ames-Test
Unpublished report (1981)

BASF AG, Abteilung Toxikologie
Report on the acute dermal irritation/corrosivity to the intact dorsal skin of the white rabbit based on OECD – Golpanol BH fest
Unpublished report (1986a)

BASF AG, Abteilung Toxikologie
Report on the acute irritation to the eye of the white rabbit based on OECD – Golpanol BH fest
Unpublished report (1986b)

BASF AG
DIN-Safety data sheet Butin-2-diol-1,4 rein krist. (1987a)

BASF AG
Written communication to BG Chemie
(1987b)

BASF AG, Abteilung Toxikologie
Test study of the oral toxicity of 2-butyne-1,4-diol (2-Butin-1,4-diol) in rats – 5 administrations by gavage in Aqua bidest., BG No.: 117
Unpublished report no. 11C0338/91040 (1992a)
On behalf of BG Chemie

BASF AG, Abteilung Toxikologie
Preliminary study of the prenatal toxicity of 2-butyne-1,4-diol in rats after oral administration (gavage), BG No.: 117
Unpublished report no. 10R0338/91057 (1992 b)
On behalf of BG Chemie

BASF AG
AIDA-Grunddatensatz 2-Butyne-1,4-diol (1993)

BASF AG, Abteilung Toxikologie
Study of the prenatal toxicity of 2-butyne-1,4-diol (2-Butin-1,4-diol) in rats after oral administration (gavage), BG No.: 117
Unpublished report no. 30R0338/91101 (1994)
On behalf of BG Chemie

Carlson, W.W., Morgan, C.C.
Effect of glycol mesylates on mouse sarcoma 180
Proc. Soc. Exp. Biol. Med., 85, 211–213 (1954)

CCR (Cytotest Cell Research GmbH & Co. KG)
Chromosome aberration assay in Chinese hamster V79 cells in vitro with Butindiol (BG-No. 117)
Unpublished report, CCR project 137406 (1989)
On behalf of BG Chemie

CCR (Cytotest Cell Research GmbH & Co. KG)
Chromosome aberration assay in Chinese hamster V79 cells in vitro with Butindiol (BG-No. 117)

Unpublished report, CCR project 137406, first amendment (1991)
On behalf of BG Chemie

Falbe, J., Regitz, M. (eds.)
Römpp Chemie Lexikon
9th ed., vol. 1, p. 532
Georg Thieme Verlag, Stuttgart, New York (1989)

Fliege, W., Voges, D., Steffan, G.
Butan-, Buten- und Butindiole
In: Ullmanns Encyklopädie der technischen Chemie
4th ed., vol.9, 19–24
Verlag Chemie, Weinheim (1975)

Gräfje, H., König, W., Weitz, H.M., Rei, W., Steffan, G., Diehl, H., Bosche, H., Schneider, K., Kieczka, H.
Butanediols, butenediol, and butynediol
In: Ullmann's encyclopedia of industrial chemistry
5th ed., vol. A4, p. 455–462
VCH Verlagsgesellschaft, Weinheim (1985)

Hoechst AG, Pharma Forschung Toxikologie und Pathologie
Butindiol – Prüfung der akuten dermalen Toxizität an der Wistar-Ratte
Unpublished report no. 88.1399 (1988)
On behalf of BG Chemie

Hoechst AG, Pharma Forschung Toxikologie und Pathologie
Butindiol – Prüfung der akuten dermalen Toxizität an der Wistar-Ratte
Unpublished report no. 90.0576 (1990)
On behalf of BG Chemie

Hüls (Chemische Werke Hüls)
Prüfung auf hautsensibilisierende Wirkung am Meerschweinchen von 1,4-Butindiol fest
Unpublished report no. 0357 (1985)

Izmerov, N.F.
Toxicometric parameters of industrial toxic chemicals under single exposure
Moscow (1982)
Cited from: RTECS (1987)

Jedrychowski, R.A., Czajkowska, T., Stetkiewicz, J., Stetkiewicz, I.
Acute toxicity of 2-butyne-1,4-diol in laboratory animals
J. Appl. Toxicol., 12 (2), 113–115 (1992a)

Jedrychowski, R.A., Czajkowska, T., Gorny, R., Stetkiewicz, J., Stetkiewicz, I.
Subacute oral toxicity of 2-butyne-1,4-diol in rats
J. Appl. Toxicol., 12 (2), 117–122 (1992b)

Knyshova, S.P.
Biological effect and hygienic significance of 1,4-butynediol and 1,4-butanediol
Hygiene and Sanitation, 33, 41–47 (1968)

Komsta, E., Secours, V.E., Chu, I., Valli, V.E., Morris, R., Harrison, J., Baranowski, E., Villeneuve, D.C.
Short-term toxicity of nine industrial chemicals
Bull. Environ. Contam. Toxicol., 43, 87–94 (1989)

Malten, K.E.
But-2-yne-1,4-diol, primary gloss improver and contact sensitizer in a nickel plating bath
Contact Dermatitis, 6, 286–287 (1980)

RCC (Research & Consulting Company AG)
Contact hypersensitivity to butin-2-diol-1,4 in albino guinea pigs, Maximization-test
Unpublished report, RCC project 263970 (1990)
On behalf of BG Chemie

Roe, F.J.C.
Tumor initiation in mouse skin by certain esters of methanesulfonic acid
Cancer Res., 17, 64–70 (1957)

RTECS (Registry of Toxic Effects of Chemical Substances)
2-Butyne-1,4-diol, ES0525000
Produced by NIOSH (National Institute for Occupational Safety and Health) (1987)

Sax, N.I. (ed.)
Dangerous properties of industrial materials, 5th ed.
van Nostrand Reinhold Company (1979)

Sprince, H., Josephs, J.A., jr., Wilpizeski, C.R.
Neuropharmacological effects of 1,4-butanediol and related congeners compared with those of gamma-hydroxybutyrate and gamma-butyrolactone
Life Sci., 5, 2041–2052 (1966)

Stasenkova, K.P., Kochetkova, T.A.
Toxikologie von 1,4-Butindiol (German translation of the Russian)
Toksikol. Novykh. Prom. Khim. Veshchestv., 7, 13–27 (1965)
Also available as an English abstract: Toxicology of 1,4-butynediol
In: Chem. Abstr., 63, 8946–8947 (1965)

Taberner, P.V., Pearce, M.J.
Hypothermic and toxic actions of 2-butyne-1,4-diol and other related diols in the rat
J. Pharm. Pharmacol., 26, 597–604 (1974)

Hydrazobenzene

1. Summary and assessment

After oral administration to rats, hydrazobenzene is excreted in the urine, both unchanged and in the form of metabolites. Benzidine, two hydroxy-derivatives of benzidine, aniline and two further unidentified compounds have been detected as metabolites. In addition, o- and p-aminophenol appear in the urine after intraperitoneal injection. No quantitative data are available.

In preliminary 4-week feeding studies, it has been shown that hydrazobenzene causes deaths in mice and rats at concentrations of 1000 mg/kg diet and above.

Hydrazobenzene is mutagenic in the Salmonella/microsome test in strain TA 100 after metabolic activation, while results in TA 98, TA 1535, TA 1537, TA 1538 and in *Escherichia coli* have been mainly negative. With metabolic activation, hydrazobenzene induces chromosome aberrations in Chinese hamster ovary (CHO) cells and increases the sister chromatid exchange rate. Furthermore, hydrazobenzene inhibits testicular DNA synthesis in mice. Hydrazobenzene has no mutagenic activity in the sex-linked recessive lethal test in *Drosophila melanogaster.*

In a National Cancer Institute (NCI) carcinogenicity study in rats and mice, technical grade hydrazobenzene was carcinogenic after oral administration in rats of both sexes and in female mice. In male rats, significantly increased numbers of hepatocellular carcinomas and Zymbal gland carcinomas occurred dose-dependently. Neoplastic nodules were observed in the liver and adenocarcinomas in the mammary glands of female rats. Female mice developed hepatocellular carcinomas, the increase in the incidence of this tumour also being statistically significant. Increases in tumour incidence have also been observed in (mainly inadequately documented) carcinogenicity studies in rats and mice involving subcutaneous and dermal application and, in strain A mice, intraperitoneal injection. Increases in Zymbal gland carcinomas, which occur very rarely spontaneously, were also seen in rats treated subcutaneously in these studies (Baker et al., 1979).

The technical grade substances used show considerable impurities and may contain benzidine as a contaminant. In 1 of 16 samples from a manufacturer's continuous production, 25 µg benzidine/g hydrazobenzene were found (Ciba-Geigy, 1982) and analysis of a residual sample of the hydrazobenzene used in the NCI study (1978) gave a benzidine content of 40 to 50 µg/g. A further analysis of 10 production lots gave an average value of 74 ± 11 µg benzidine/g hydrazobenzene, a value which is considered to be too high because of the corresponding aniline content (Hüls, 1989 c). Benzidine is carcinogenic in man. It was first included in the list of MAK values in 1966 and has been classified in category III A 1 since 1975 and in EC category C1 in TRGS 500 since 1993 (DFG, 1975; TRGS 500, 1993). Benzidine can be formed by the metabolism of hydrazobenzene.

In a study of a group of 189 benzidine production workers who were occupationally exposed to hydrazobenzene, 20 cases of bladder cancer were observed (10.6%). These are, however, more likely to be attributable to benzidine.

Within the Employment Accident Insurance Fund of the Chemical Industry (BG Chemie), development of a bladder tumour following exposure to hydrazobenzene has been recognized as an occupational disease under § 551 paragraph 2 of the Federal Insurance Order.

In the past, isolated cases of irritant cystitis with incontinence and haematuria have been seen on exposure to high concentrations of hydrazobenzene.

2. Name of substance

2.1	Usual name	Hydrazobenzene
2.2	IUPAC-name	1,2-Diphenylhydrazine
2.3	CAS-No.	122–66–7
2.4	EINECS-No.	204–563–5

3. Synonyms, common and trade names

N,N'-Bianiline
N,N'-Diphenylhydrazine
1,1'-Hydrazobisbenzene
Hydrazodibenzene
sym-Diphenylhydrazine
Hydrazobenzol

4. Structural and molecular formulae

4.1	Structural formula	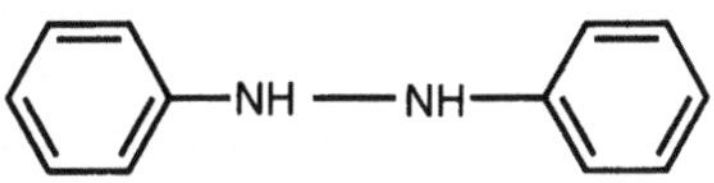
4.2	Molecular formula	$C_{12}H_{12}N_2$

5. Physical and chemical properties

5.1	Molecular mass, g/mol	184.24
5.2	Melting point, °C	ca. 125 (Hüls, 1989 a)
5.3	Boiling point, °C	No information available
5.4	Density, g/cm^3	1.16 (at 20 °C) (Hüls, 1989 a)
5.5	Vapour pressure, hPa	No information available
5.6	Solubility in water	< 0.01 g/l (at 20 °C) (Hüls, 1989 a)
5.7	Solubility in organic solvents	Dissolves readily in alcohol, ether (Falbe and Regitz, 1990)
5.8	Solubility in fat	No information available
5.9	pH value	No information available
5.10	Conversion factor	1 ml/m^3 (ppm) ≙ 7.53 mg/m^3 1 mg/m^3 ≙ 0.133 ml/m^3(ppm) (at 1013 hPa and 25 °C)

6. Uses

Intermediate in the manufacture of pharmaceuticals (Hüls, 1989b).

7. Experimental results

7.1 Toxicokinetics and metabolism

After oral administration of 200 or 400 mg hydrazobenzene/kg body weight to rats, hydrazobenzene, benzidine, aniline, two hydroxy derivatives of benzidine and two further unidentified metabolites were found in the urine by means of thin layer chromatography. Hydrazobenzene, aniline, benzidine, p-aminophenol and o-aminophenol were detected in the urine after intraperitoneal administration of 100 or 200 mg/kg body weight. Following intravenous or intratracheal administration, only one metabolite, which appeared to be of a phenolic nature, was detectable in the urine (no further details; Dutkiewicz and Szymanska, 1973).

Aniline was detectable as a metabolite after hydrazobenzene was incubated with rat gut microorganisms at 37 °C for 7 days under anaerobic conditions (no quantitative data provided; Bolton and Griffiths, 1978).

The formation of benzidine from hydrazobenzene was detectable in acid solution in in vitro studies (see Sect. 8; Mappes, 1987).

7.2 Acute and subacute toxicity

In order to determine the maximum tolerable concentration for use in carcinogenicity studies, 4-week feeding studies with a 2-week observation period were carried out in groups of 5 male and 5 female Fischer-344 rats and B6C3F1 mice. The concentrations of technical-grade hydrazobenzene tested in male rats and mice were 70, 140, 280, 550, 1080, 2140, 3010 and 4230 mg/kg diet (equivalent to about 10.5, 21.0, 42.0, 82.5, 162.0, 321.0, 451.5, and 634.5 mg/kg body weight in the rats and about 4.9, 9.8, 19.6, 38.5, 75.6, 151.2, 210.7 and 296.1 mg/kg body weight in the mice), while in female rats and mice 0.8 (rats only), 3, 11, 20, 40, 150, 1040, 7310 and 51 380 mg/kg diet were tested (equivalent to about 0.12,

0.45, 1.65, 3.0, 6.0, 22.5, 156.0, 1096.5 and 7707.0 mg/kg body weight in rats and about 0.21, 0.77, 1.4, 2.8, 10.5, 72.8, 511.7 and 3597 mg/kg body weight in mice). Corresponding control groups were studied simultaneously. Of the male rats in the 1080 mg/kg diet group, 2/5 died, while at higher concentrations, all male and female rats died. No treatment-related pathological effects were found. Of the male mice, 1/5 in the 3010 mg/kg diet group and 2/5 in the 4230 mg/kg diet group died. In the 7310 mg/kg diet group 4/5 females died and all 5 females in the 51,380 mg/kg diet group died. The only pathological finding in the mice was bleeding in the gastrointestinal tract (NCI, 1978).

7.3 Skin and mucous membrane effects

No information available.

7.4 Sensitization

No information available.

7.5 Subchronic and chronic toxicity

The following non-neoplastic effects were found in carcinogenicity studies in rats and mice (study protocol described in Sect. 7.7 (NCI, 1978)):

A slight decrease in body weight was seen in the high dose male rats (300 mg/kg diet) during the course of the study. Female rats showed a slight decrease in body weight in the low dose group (40 mg/kg diet) after 46 weeks, and in the high dose group (100 mg/kg diet) after 22 weeks. In the male rats, there was no difference in survival between the treated and untreated animals, while in the female rats, survival in the high dose group was significantly reduced relative to the controls (Cox test, $p < 0.001$). In week 100 of the study, 25/50 of the females in the high dose group (100 mg/kg diet) and 43/50 in the control group were still alive. No decrease in survival rate was seen in the low dose group (40 mg/kg diet).

In the mice, there were no changes in body weight compared with the controls in the low dose groups (70 to 80 mg/kg diet in males, 40 mg/kg diet in females). In males and females in the high dose group (400 mg/kg diet) body weight was reduced to 20 to 25 g after week 28 (controls 35 to 45 g). No other prominent clinical effects were found. Survival was significantly lower in male and female mice in the high dose group than in the corresponding controls (33/50 and 26/50 compared with 39/50 and 38/50). No reduction in the survival rate was seen in the low dose group.

7.6 Genotoxicity

7.6.1 In vitro

Technical grade hydrazobenzene (purity unspecified) was studied in the preincubation test in the *Salmonella typhimurium* strains TA 1535, TA 1537, TA 98 and TA 100 with and without metabolic activation (S9-mix from Aroclor 1254-induced livers from male Sprague-Dawley rats and male Syrian hamsters). The concentrations tested ranged from 3.3 to 333.3 µg/plate. Hydrazobenzene was only mutagenic in strain TA 100 from 33.3 µg/plate after metabolic activation. No mutagenic activity was evident in the other strains with or without metabolic activation (Haworth et al., 1983).

In a ring study in 4 independent institutes, hydrazobenzene (at least 65% pure, no further details on contaminants) was studied, along with 62 other substances, in a plate incorporation test in the *Salmonella typhimurium* test strains TA 98, TA 100, TA 1535, TA 1537 and TA 1538 and in *Escherichia coli* WP2uvrA. The test was carried out with and without metabolic activation (S9-mix from Aroclor 1254-induced livers from rats, mice and hamsters) and the concentrations tested ranged from 0.1 to 6666.6 µg/plate. Hydrazobenzene was also mutagenic in this study in TA 100 at concentrations of from 10 to 333.3 µg/plate following metabolic activation. The results in the other strains were predominantly negative and those in *Escherichia coli* were unequivocally negative (Dunkel et al., 1985).

The ability of hydrazobenzene to induce chromosome aberrations and affect the sister chromatid exchange rate was studied in Chinese hamster ovary (CHO-W-B1) cells in vitro with and without metabolic activation (S9-mix from Aroclor 1254-induced rat liver). The concentrations used in the chromosome aberration test were 1.4 to 42 µg/ml. A questionable result was obtained without metabolic activation at the highest concentration (42 µg/ml), while with metabolic activation, concentrations of from 14 to 42 µg/ml showed a significant positive result ($p \leq 0.05$). In the SCE test, the concentrations tested were between 1.4 and 15 µg/ml. Hydrazobenzene was negative without metabolic activation up to the highest concentration (14 µg/ml), while with metabolic activation, it was positive ($p < 0.05$) in one test at the highest concentration (14 µg/ml) and in a second test at all concentrations (5, 10 and 15 µg/ml). Thus in these studies, hydrazobenzene caused chromosome aberrations and increased the sister chromatid exchange rate (Galloway et al., 1987).

7.6.2 In vivo

The effect of hydrazobenzene was also studied in a short-term test for the inhibition of testicular DNA synthesis by potential mutagens and carcinogens in mice. Male mice were given an intraperitoneal injection of 100 mg hydrazobenzene/kg body weight 3.5 h before they were killed. ^3H-thymidine was administered by intraperitoneal injection 30 minutes before sacrifice and its incorporation into testicular DNA was measured. Hydrazobenzene caused a 72% decrease in thymidine incorporation ($p < 0.01$; Seiler, 1977).

Hydrazobenzene was studied in the sex-linked recessive lethal test in *Drosophila melanogaster.* No increase in lethal mutations occurred after feeding 50 ppm for 3 days or following injection of 80 ppm (no further details; Yoon et al., 1985).

7.7 Carcinogenicity

In an early study, 52 female Sherman rats (ca. 150 g) were each given weekly subcutaneous injections of 60 mg of technical grade hydrazobenzene in olive oil for their entire lives (cumulative total amount maximum 6.4 g). 31 rats survived for longer than 300 days (controls 28/50). In contrast to benzidine, which was tested at the same time, no liver tumours and only one tumour of the Zymbal gland occurred after 366 days (Spitz et al., 1950).

A long-term feeding study was carried out with technical grade hydrazobenzene in Fischer-344 rats and B6C3F1 mice. Analyses of the remains of the test sample gave benzidine contents of 0.004 and

0.005% (equivalent to 40 to 50 µg/g; BIA, 1984 a, b). The study groups each consisted of 50 male and 50 female animals (except female mice, of which there were only 47 in the low dose group). The control groups consisted of 100 male and 100 female animals (except male rats, of which there were only 99). The concentrations were 70 to 80 and 300 mg/kg diet in the male rats and 40 and 100 mg/kg diet in the female rats. Male mice received 70 to 80 or 400 mg/kg diet, female mice 40 or 400 mg/kg diet. The animals were fed these diets for 78 weeks. There then followed an observation period of 28 to 30 weeks in the rats and 17 to 18 weeks in the mice. Feeding hydrazobenzene to rats increased the formation of neoplasms of the liver in both sexes, of the Zymbal gland (in the auditory canal) in males and of the mammary glands in females. Hepatocellular carcinomas were found in 5/49 males in the low dose group ($p = 0.031$), 31/49 males in the high dose group ($p < 0.001$) and 1/95 animals in the control group. No liver carcinomas were observed in the female rats, although hyperplastic nodules occurred in 6/50 of the animals in the highest dose group (controls 0/50; $p = 0.013$). Tumours of the Zymbal gland and the skin of the ear occurred in 1/95 male controls, 2/50 males in the low dose group, 7/49 males in the high dose group ($p = 0.007$), 0/98 female controls, 1/50 females in the low dose group and 2/50 females in the high dose group. Of these tumours, 8 were squamous epithelial carcinomas, 1 was a squamous epithelial papilloma and 2 were adenocarcinomas (the histology of 2 tumours was not described). Adenocarcinomas of the mammary gland occurred in 1/48 controls, 3/50 low dose females and 6/50 female rats in the high dose group ($p = 0.013$). The female mice showed an increased occurrence of hepatocellular carcinomas (20/43 in the high dose group, 4/39 in the low dose group and 3/97 in the control group). The increase in the tumour rate in the high dose group was statistically significant ($p < 0.001$). There were no indications that hydrazobenzene was carcinogenic in male mice (NCI, 1978).

The average tumorigenic dose (TD_{50}) for the individual types of tumours was later calculated by other authors for the NCI study. These were, for the Zymbal tumours (male rats) 38.5 mg/kg body weight/day, for liver tumours (female mice and male rats) 26.0 and 3.70 mg/kg body weight/day, respectively, and for mammary gland tumours (female rats) 11.40 mg/kg body weight/day. For all target organs together, the value for male rats was 3.55 mg/kg body weight/day (Gold et al., 1986).

In a long-term study in 91 male and female rats (strain unspecified), hydrazobenzene was injected subcutaneously once weekly, initially at a dose of 60 mg/rat and then at 40 mg/rat (equivalent to ca. 300 and 200 mg/kg body weight, respectively) in 0.3 and 0.5 ml sunflower oil, respectively. The injection period was 588 days, the total dose during this time was 3.8 g hydrazobenzene per animal. The first neoplasm (a Zymbal gland carcinoma) occurred on day 187 of the study after the animals had received 1.3 g hydrazobenzene. At this time, 53 rats were still alive (29 males and 24 females). A further 11 tumours developed before the end of the study, which lasted 692 days (1 uterine adenocarcinoma, 2 mammary carcinomas, 1 mammary adenoma, 2 Zymbal gland tumours, 2 liver adenomas, 1 spleen haemangioendothelioma, 1 reticulosis). No information was given about a control group (Pliss, 1974).

In a further study by the same author, a total of 72 male and female rats (strain not specified) were given ca. 150 mg/kg body weight hydrazobenzene daily, which was dissolved in 0.5 ml sunflower oil and added to their feed. The substance was administered 5 times per week over 590 days, the feed of each rat having a total of 12.57 g hydrazobenzene added to it. Tumours were found in 21 rats with an average latent period of 566 days (5 Zymbal gland tumours, 6 liver tumours, of which 4 were adenomas and 2 were haemangiomas, 4 mammary gland tumours, 3 uterine carcinomas, 2 kidney tumours, 1 leucosis). Information on a control group was also not provided in this study (Pliss, 1974).

The same author also tested hydrazobenzene in mice. Once a week, 60 mice were each given a subcutaneous injection of 5 mg hydrazobenzene in 0.2 ml sunflower oil (ca. 200 mg/kg body weight) for 512 days (ca. 73 weeks). The total dose per mouse was 370 mg. The first tumour (a sarcoma at the injection site) occurred on day 265. At this time, 30 mice (29 males and 1 female) were still alive. After 365 days 22 mice were still in the study, while after 548 days only 3 remained. In total, tumours developed in 11 mice (2 sarcomas at the injection site, 5 lung adenomas, 3 liver tumours, 1 leucosis). In the control group, spontaneous tumours were said to have occurred in 17% of the animals, but no further details were given about this observation (Pliss, 1974).

In a further study by the same author, hydrazobenzene dissolved in benzene was applied dermally to 50 mice (2 mg/mouse in 0.05 ml benzene, 3 times weekly for 442 days). A total of 360 mg hydrazobenzene was applied. No effects were evident at the application site (between the shoulder blades). The first tumour (a skin tumour on the hind leg) was observed on day 179 of the study. By this time, 12 female and 29 male mice were still alive. After 1 year, 3 females and 13 males were still alive and after 1.5 years, only 4 males survived. In 9 animals, 1 skin cancer, 2 haemangiomas of the liver, 3 leukoses and 3 lung adenomas occurred. The control group was apparently the same as that used in the study described above, in which 17% of the animals developed tumours (no further details; Pliss, 1974).

A conspicuous feature of these studies is the high loss of animals within the period before the occurrence of the first tumour. According to Pliss, the high loss is both a consequence of infection and is treatment-related. However, further information on this assertion was not provided. On the whole, the studies of Pliss are inadequately reported.

Another long-term study with technical grade hydrazobenzene was carried out in 50 rats (25 males and 25 females, strain not specified). The test substance was collected from the production plant and consisted of 55% hydrazobenzene. The animals were injected subcutaneously once weekly with 20 mg of this substance in 0.5 ml sunflower oil. In 1 year, 50 injections were administered. The rats were then observed until their natural death or until they died due to tumour development. When the first tumour occurred, 19 rats were still alive, 4 of which developed tumours (1 sarcoma at the injection site, 1 Zymbal gland carcinoma, 1 kidney carcinoma and 1 mucous membrane hyperplasia of the endometrium). The average latent period was 603 ± 45 days. In a control group of the same size, 1 mammary fibroadenoma was observed after 630 days in 36 surviving animals. With the same study protocol and the same control group, the authors tested a mixture of substances (27.2% benzidine sulfate) which contained benzidine (produced by chemical rearrangement of hydrazobenzene). The rats were injected with 15 mg of this mixture subcutaneously once a week. After an average latent period of 599 ± 39 days, 5 tumours were diagnosed in 14 surviving animals

(1 sarcoma at the injection site, 3 Zymbal gland carcinomas, 1 carcinometastasis of unclear origin; Kurlyandsky et al., 1976).

On simultaneous subcutaneous administration of benzidine- and hydrazobenzene-containing mixtures (15 mg and 20 mg/animal, same study protocol as previous study, same control group), 10 tumours were found in the 14 surviving animals after an average latent period of 506 ± 15 days (4 sarcomas at the injection site, 2 Zymbal gland carcinomas, 1 hepatoma, 1 mammary carcinoma and 2 mammary fibroadenomas). The animals in the control group were injected with 0.5 ml sunflower oil. In this group, 1 mammary fibroadenoma occurred in 26 surviving animals. The authors concluded that the simultaneous administration of benzidine and hydrazobenzene led to a higher tumour rate (71.4%) than administration of the individual substances (35.7% and 21%, respectively). However, it should be taken into account that the test substances were not available in pure form, but were mixtures of substances (Genin et al., 1975; Kurlyandsky et al., 1976).

The ability of hydrazobenzene (purity unspecified) to induce lung tumours was studied in a screening test in strain A mice, which are particularly susceptible to these tumours (spontaneous incidence begins at the age of 3 to 4 months and is nearly 100% after 24 months). Groups of 10 male and 10 female mice were given intraperitoneal injections of 200, 100 or 50 mg/kg body weight (in tricaprylin) three times a week over a period of 8 weeks. The animals then remained untreated for a further 16 weeks. The tumours were evaluated after a total of 24 weeks. There was a significant increase in the lung tumour rate (alveolar and bronchial adenomas) in male mice at the highest dose level (200 mg/kg body weight) in terms of the percentage of tumour-carrying survivors (67%; solvent controls 13%) and the number of tumours/mouse (0.89, $p \leq 0.05$; solvent control 0.167). In the females, only the number of tumours per surviving mouse (0.56; solvent controls 0.11) was significantly increased ($p \leq 0.05$) at the same dose level (Maronpot et al., 1986).

7.8 Reproductive toxicity

No information available.

7.9 Effects on the immune system

No information available.

7.10 Neurotoxicity

No information available.

7.11 Other effects

No information available.

8. Experience in humans

From 1940 to 1960, benzidine was manufactured in the East Bohemian Chemical Works under relatively poor conditions of hygiene, in the course of which hydrazobenzene arose as an intermediate. In 1960, the business was moved to a new, modern building. The workplace concentrations of hydrazobenzene measured there from 1961 to 1965 were between 0.006 and 3 µg/l air and were thus below the maximum permissible standard in the former USSR of 10 µg/l air. Out of a total of 329 people exposed since 1940, 189 were available for examination. Of these, bladder carcinoma was established in 20 cases (10.6%). The average period of exposure was 8 years (Kuzelová et al., 1969). It is however, very unlikely that the bladder carcinomas were caused by hydrazobenzene per se, and benzidine was probably the causative agent.

In 1951, a case of papillomatosis of the urinary bladder was reported in a 47-year-old patient who had been occupationally exposed to trinitrotoluene for 6 years and then benzidine and hydrazobenzene for 3 $^{3}/_{4}$ years. He had also had contact with other intermediates to a lesser extent. At the time of diagnosis, he had not been exposed for 1 year (Marhold, 1951). The causative involvement of hydrazobenzene is not very probable.

In earlier years, isolated cases of irritant cystitis with incontinence and haematuria together with a feeling of pressure in the region of the bladder and kidneys were observed after exposure to high concentrations of hydrazobenzene. Bladder cancers have not been seen on contact with hydrazobenzene (Ciba-Geigy, 1985).

It has also been reported that no indication of hydrazobenzene having carcinogenic activity in humans was found in documentation going back to 1914 in a chemical plant (Korallus, 1985, 1989).

A carcinoma of the urinary bladder developed in a 59-year-old chemist who, 10 years previously, had been exposed to hydrazobenzene dust for 2 to 3 h daily for about 3 years. An expert opinion prepared in connection with the illness was based on in vitro studies, which were carried out under conditions similar to those prevailing in the stomach. It was determined that, from every 10 mg hydrazobenzene, 3.6, 3.2 and 2.0 mg benzidine were formed in 2 h in the presence of 0.07 n hydrochloric acid, bisulfate buffer (pH 2) and bisulfate buffer (pH 2.5), respectively. Based on these results it was assumed that benzidine formed from hydrazobenzene under the influence of stomach acid was, to a considerable extent, responsible for the development of the bladder tumour (Mappes, 1987).

The above-mentioned illness was recognized as an occupational disease by BG Chemie under § 551, paragraph 2 of the Reichsversicherungsordnung (BG Chemie, 1990).

9. Threshold limit values

Hydrazobenzene is classified as C2 (the substance has to be considered carcinogenic in humans) (TRGS, 1994).

References

Auterhoff, H., Pauli, H.P.
Zur Kenntnis der Identitätsreaktionen des Phenylbutazons
Arch. Pharm., 309, 538–542 (1976)

Baker, H.J., Lindsey, J.R., Weisbroth, S.H. (eds.)
The laboratory rat
Vol. I, p. 337
Academic Press, New York (1979)

BG Chemie (Berufsgenossenschaft der chemischen Industrie)
Internal communication from the regional management in Cologne to head office (1990)

BIA (Berufsgenossenschaftliches Institut für Arbeitssicherheit, St. Augustin)
Written communication to BG Chemie of 23.02.1984 a

BIA (Berufsgenossenschaftliches Institut für Arbeitssicherheit, St. Augustin)
Written communication to BG Chemie of 13.03.1984 b

Bolton, G.C., Griffiths, L.A.
Metabolism of hydrazines and hydrazides by the intestinal microflora
Experientia, 34 (11), 1484–1486 (1978)

Ciba-Geigy AG, Werk Grenzach, Werksärztliche Abteilung
Written communication to BG Chemie of 29.12.1982

Ciba-Geigy AG, Werk Grenzach, Werksärztliche Abteilung
Written communication to BG Chemie of 15.07.1985
DFG (Deutsche Forschungsgemeinschaft, Senatskommission zur Prüfung gesundheitsschädlicher Arbeitsstoffe)
Mitteilung XI, MAK-Werte-Liste 1975
Harald Boldt Verlag, Boppard (1975)

Dunkel, V.C., Zeiger, E., Brusick, D., McCoy, E., McGregor, D., Mortelmans, K., Rosenkranz, H.S., Simmon, V.F.
Reproducibility of microbial mutagenicity assays: II. Testing of carcinogens and noncarcinogens in *Salmonella typhimurium* and *Escherichia coli*
Environ. Mutagen., 7, Suppl. 5, 1–248 (1985)

Dutkiewicz, T., Szymanska, J.
Chromatographic determination of hydrazobenzene metabolites in rats (Polish work, available only as an abstract)
Bromatol. Chem. Toksykol., 6, 323–327 (1973)

Falbe, J., Regitz, M. (eds.)
Römpp Chemie Lexikon
9th ed., vol.3, p. 1880
Georg Thieme Verlag, Stuttgart (1990)

Galloway, S.M., Armstrong, M.J., Reuben, C., Colman, S., Brown, B., Cannon, C., Bloom, A.D., Nakamura, F., Ahmed, M., Duk, S., Rimpo, J., Margolin, B.H., Resnick, M.A., Anderson, B., Zeiger, E.
Chromosome aberrations and sister chromatid exchanges in Chinese hamster ovary cells: evaluations of 108 chemicals
Environ. Mol. Mutagen., 10, Suppl. 10, 1–175 (1987)

Genin, V.A., Medvedovsky, A.G., Voronin, V.M.
Über die Verstärkung der kanzerogenen Aktivität bei der gleichzeitigen Wirkung von Hydrazobenzol und Benzidinsulfat (German translation of the Russian)
Gig. Tr. Prof. Zabol., 4, 28–31 (1975)

Gold, L.S., Ward, J.M., Bernstein, L., Stern, B.
Association between carcinogenic potency and tumor pathology in rodent carcinogenesis bioassays
Fundam. Appl. Toxicol., 6, 677–690 (1986)

Haworth, S., Lawlor, T., Mortelmans, K., Speck, W., Zeiger, E.
Salmonella mutagenicity test results for 250 chemicals
Environ. Mutagen., 5, Suppl. 1, 3–142 (1983)

Hüls AG, Marl
DIN-Safety data sheet "Hydrazobenzol" (1989 a)

Hüls AG, Marl
Written communication to BG Chemie of 26.04.1989 b

Hüls AG, Marl
Written communication to BG Chemie of 06.12.1989 c

Korallus, U.
Written communication to BG Chemie of 16.09.1985

Korallus, U.
Written communication to BG Chemie of 11.08.1989

Kurlyandsky, B.A., Medvedovsky, A.G., Genin, V.A., Voronin, V.M., Mashbits, F.D.
Experimental study on a combined action of some diphenyl amino-derivatives with a view to preventing occupational neoplasms of the urinary bladder (English translation of the Russian)
Gig. Tr. Prof. Zabol., 5, 34–38 (1976)

Kuzelová, M., Kunor, V., Hurt, K.
Überwachung von bei der Benzidinherstellung beschäftigten Arbeitskräften (German translation of the Czech)
Pracovni Lékarstvi, 21, vol. 7, 310–313 (1969)

Mappes, R.
Written communication to the regional management of BG Chemie in Cologne of 17.07.1987

Marhold, J.
A case of tumour of the bladder following exposure to benzidine
Pracovni Lékarstvi, 3, 272–274 (1951)

Maronpot, R.R., Shimkin, M.B., Witschi, H.P., Smith, L.H., Cline, J.M.
Strain A mouse pulmonary tumor test results for chemicals previously tested in the National Cancer Institute carcinogenicity tests
J. Natl. Cancer Inst., 76, no. 6, 1101–1112 (1986)

NCI (National Cancer Institute)
Bioassay of hydrazobenzene for possible carcinogenicity
Technical Report Series no. 92, NCI-CG-TR-92 (1978)

Organikum, Organisch-chemisches Grundpraktikum
9th ed., p. 633
VEB Deutscher Verlag der Wissenschaften, Berlin (1969)

Pliss, G.B.
Über die kanzerogenen Eigenschaften von Hydrazobenzol (German translation of the Russian)
Vopr. Onkol., 20, 53–57 (1974)

Seiler, J.P.
Inhibition of testicular DNA synthesis by chemical mutagens and carcinogens. Preliminary results in the validation of a novel short term test
Mutat. Res., 46, 305–310 (1977)

Spitz, S., Maguigan, W.H., Dobriner, K.
The carcinogenic action of benzidine
Cancer, 3, 789–804 (1950)

TRGS (Technische Regel für Gefahrstoffe) 905
Bekanntmachung des Bundesministers für Arbeit und Sozialordnung
Bundesarbeitsblatt, vol 6, p. 53–63 (1994)

Yoon, J.S., Mason, J.M., Valencia, R., Woodruff, R.C., Zimmering, S.
Chemical mutagenesis testing in Drosophila. IV. Results of 45 coded compounds tested for the National Toxicology Program
Environ. Mutagen., 7, 349–367 (1985)

4-Nitrodiphenylamine

1. Summary and assessment

The acute toxicity (LD_{50}) of 4-nitrodiphenylamine by the oral route in rats and the dermal route in rabbits is low, being > 7940 mg/kg body weight.

On repeated administration of the substance in the diet, the no-effect level in the rat is 2000 ppm (equivalent to ca. 100 mg/kg body weight/day). Higher concentrations cause effects on the kidneys. Inhalation of 29 and 52 mg/m^3 for 21 days causes raised methaemoglobin levels and other haematological effects (not precisely defined), as well as increased liver weights. The no effect level is 8.5 mg/m^3.

4-Nitrodiphenylamine is not irritating to rabbit skin and is slightly irritating to the rabbit eye.

After subchronic oral administration in the diet (ca. 40 to 250 mg 4-nitrodiphenylamine/kg body weight/day for 90 days), the no-effect level in the rat was ca. 40 mg/kg body weight/day. The methaemoglobin level was increased at the intermediate and high dose levels, and effects on the kidneys occurred in males in the high dose group.

Two studies on the mutagenicity of 4-nitrodiphenylamine have given conflicting results in the Salmonella/microsome test. No data on the purity of the products used are available. There were no indications of damage to the chromosomes in a cytogenetic study in Chinese hamster cells (CHO cell line). The substance was negative in the UDS test and in a test for DNA replication in rat hepatocytes in vivo. There were also no indications that 4-nitrodiphenylamine had genotoxic activity in the micronucleus test in the rat in vivo.

4-Nitrodiphenylamine was embryotoxic but not teratogenic in the rat after oral administration at doses that were maternally toxic (1000 and 3000 mg/kg body weight). At 250 mg/kg body weight it caused neither maternal nor embryotoxicity.

2. Name of substance

2.1	Usual name	4-Nitrodiphenylamine
2.2	IUPAC-name	4-Nitro-N-phenylbenzenamine
2.3	CAS-No.	836–30–6
2.4	EINECS-No.	212–646–2

3. Synonyms, common and trade names

Benzenamine, 4-nitro-N-phenyl
4-NDPA
p-Nitrodiphenylamine
p-Nitrophenylphenylamine
Phenyl 4-nitrophenylamine
4-Nitrodiphenylamin

4. Structural and molecular formulae

4.1 Structural formula

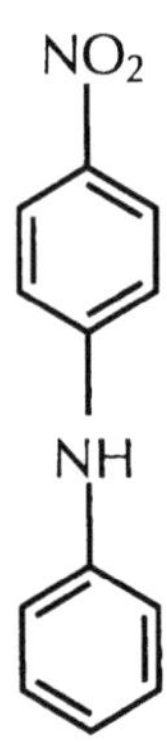

4.2 Molecular formula $C_{12}H_{10}N_2O_2$

5. Physical and chemical properties

5.1	Molecular mass, g/mol	214.24
5.2	Melting point, °C	120–130 (Bayer, 1991) 132 (Monsanto, 1979) 132–133 (Sax and Lewis, 1989) 133–134 (Weast, 1988/89)
5.3	Boiling point, °C	211 (at 26.7 hPa) (Weast, 1988/89) 343 (at 1013 hPa) (Monsanto, 1979)
5.4	Vapour pressure, hPa	< 0.1 (at 50 °C) (Bayer, 1991) 1.33 (at 192 °C) (Monsanto, 1979) 10.67 (at 240 °C) (Monsanto, 1979)
5.5	Density, g/cm^3	ca. 0.55 (bulk density) (Bayer, 1991)
5.6	Solubility in water	0.0041 g/l (at 24 °C) (Monsanto, undated)
5.7	Solubility in organic solvents	Soluble in ethanol and acetic acid (Sax and Lewis, 1989; Weast, 1988/89)
5.8	Solubility in fat	n-Octanol/water partition coefficient, log P_{ow}: 3.82 (measured at 25 °C) (Monsanto, undated)
5.9	pH value	ca. 11 (at 100 g/l) (Bayer, 1990)
5.10	Conversion factor	1 ml/m^3 (ppm) ≙ 8.74 mg/m^3 1 mg/m^3 ≙ 0.11 ml/m^3 (ppm) (at 1013 hPa and 25 °C)

6. Uses

Intermediate in the manufacture of the rubber antioxidant N-(1,3-dimethylbutyl)-N′-phenyl-p-phenylenediamine (Bayer, 1993).

7. Experimental results

7.1 Toxicokinetics and metabolism

No information available.

7.2 Acute and subacute toxicity

To determine the acute oral toxicity of 4-nitrodiphenylamine (ca. 94% pure), male and female Sprague-Dawley rats (n = 5) received a single dose of 7940 mg/kg body weight as a 25% suspension in corn oil by gavage. The observation period was 7 days. None of the rats died and the LD_{50} was thus > 7940 mg/kg body weight. Signs of toxicity included anorexia and reduced activity during the first 1 to 2 days after administration. No macroscopic effects were observed (on the viscera; no further details; Randall and Bannister, 1990).

A further LD_{50} value of 11,800 mg/kg body weight was reported for the rat following the administration of a single oral dose of 4-nitrodiphenylamine (no further details; Marhold, 1972).

To determine the acute dermal toxicity, 4-nitrodiphenylamine (ca. 94% pure) was applied to the clipped intact skin of New Zealand white rabbits (n = 1 and 2) under semi-occlusive cover as a 40% suspension in corn oil at doses of 5010 and 7940 mg/kg body weight, respectively, for 24 h. The observation period was 14 days. No deaths occurred and the LD_{50} was thus > 7940 mg/kg body weight. Signs of toxicity included anorexia and reduced activity during the first 2 to 4 days after application. No macroscopic effects (no further details) were observed at autopsy (Randall and Bannister, 1990).

Groups of 5 male and 5 female Sprague-Dawley rats were given 4-nitrodiphenylamine in the diet for 4 weeks at concentrations of 0, 2000, 5000, 10,000 and 20,000 ppm (equivalent to ca. 100, 250, 500 and 1000 mg/kg body weight/day at an assumed feed intake of 50 g/kg body weight/day). Reductions in body weight gain accompanying reduced feed intake were seen in males and females at the two highest dose levels. The kidneys of the rats given 5000, 10,000 and 20,000 ppm had pigments and casts in the tubules, and cysts were also evident in these animals. The no effect level was given as 2000 ppm (no further details; Monsanto, 1983 a).

Groups of 10 Sprague-Dawley rats/sex inhaled 8.5, 29 or 52 mg 4-nitrodiphenylamine/m^3 for 6 h/day, 5 days/week for 21 days (whole body exposure). Increased methaemoglobin levels and other haematological effects (of an unspecified nature) were observed in the rats exposed to 29 or 52 mg/m^3, together with increases in the absolute and relative liver weights. Irritant effects in the anterior nasal cavity were reported at the highest concentration. The no effect level was given as 8.5 mg/m^3 (no further details; Monsanto, 1983 b).

7.3 Skin and mucous membrane effects

The irritant effect of 4-nitrodiphenylamine (ca. 94% pure) was tested on the clipped skin of a total of 6 male and female New Zealand white rabbits under semi-occlusive conditions. The rabbits were treated with 500 mg finely ground 4-nitrodiphenylamine, which was applied once to the intact or scarified skin as an aqueous preparation for 24 h. The observation period was 7 days and the skin reaction was evaluated in accordance with the Federal Hazardous Substances Act, 21 CFR, § 191.11. The primary irritation index was 0 (maximum irritation index 8). Thus the substance was not irritating to the skin of the rabbit (Randall and Bannister, 1990).

To study the irritant effect of the substance on the eye, a total of 6 male and female New Zealand white rabbits each had 100 mg of finely ground 4-nitrodiphenylamine (ca. 94% pure) placed in the conjunctival sac of one eye. The observation period was 7 days and the eye irritation was evaluated in accordance with the Federal Hazardous Substances Act, 21 CFR, § 191.12. The irritation index was 2.6 (maximum irritation index 110). The substance was evaluated as slightly irritating to the eye of the rabbit (Randall and Bannister, 1990).

7.4 Sensitization

No information available.

7.5 Subchronic and chronic toxicity

Groups of 30 male and 30 female Sprague-Dawley rats were given 4-nitrodiphenylamine in the diet at concentrations of 0 (control), 800, 2000 and 5000 ppm (equivalent to ca. 40, 100 and 250 mg/kg body weight/day at an assumed feed intake of 50 g/kg body weight/day) for 90 days. Body weight was reduced in the high dose group and the methaemoglobin level was increased from 2000 ppm. Effects on the kidneys occurred in the males in the high dose group. The no effect level was given as ca. 40 mg/kg body weight/day (no further details; Monsanto, 1983 c).

7.6 Genotoxicity

7.6.1 In vitro

4-Nitrodiphenylamine (86.4% pure) was tested for mutagenic activity in the *Salmonella typhimurium* strains TA 98, TA 100, TA 1535 and TA 1537 at concentrations of from 1 to 1000 µg/plate in the standard plate test with and without metabolic activation (S9-mix from Aroclor 1254-induced livers from male rats). At each concentration and for each strain, 3 plates were used. The study was repeated in an independent experiment. Cytotoxicity was observed at concentrations of 333.3 µg/plate and above. A significant and reproducible dose-dependent increase in the number of revertant colonies was seen in strains TA 98 and TA 1537 both with and without S9-mix, with the induction of mutants in strain TA 98 described as relatively weak without metabolic activation. The test substance did not induce gene mutations in strains TA 100 or TA 1535 either with or without metabolic activation. Overall, the substance was evaluated as mutagenic in this test system (CCR, 1989).

In contrast, in a further Ames test in the *Salmonella typhimurium* strains TA 98, TA 100, TA 1535, TA 1537 and TA 1538, there were no indications that 4-nitrodiphenylamine was mutagenic either with or without metabolic activation. The substance was tested at concentrations of from 5 to 500 µg/plate. Without metabolic activation, 4-nitrodiphenylamine was cytotoxic from 167 µg/plate (no further details; Monsanto, 1986 a).

There were also no indications that 4-nitrodiphenylamine was mutagenic with or without metabolic activation in a chromosome aberration test in Chinese hamster ovary (CHO) cells. The chemical was tested at concentrations of from 3 to 40 µg/ml without metabolic activation and from 5 to 25 µg/ml with metabolic activation. Concentrations of 50 µg/ml with metabolic activation and 30 µg/ml without metabolic activation were cytotoxic (no further details; Monsanto, 1986 b).

7.6.2 In vivo

In the micronucleus test in Sprague-Dawley rats, there were no indications of mutagenic activity following a single dose of 5000 mg 4-nitrodiphenylamine/kg body weight (route unspecified). The bone marrow was processed 6, 18 and 30 h after dosing (no further details; Monsanto, 1986 c).

Primary hepatocytes were isolated from groups of Sprague-Dawley rats 16 h (UDS test) or 48 h (DNA replication) after administration of single doses of 0, 50, 250 or 1000 mg 4-nitrodiphenylamine/kg body weight by gavage. In this study (which was carried out in accordance with the method of Mirsalis, J.C. and Butterworth, B.E., Carcinogenesis, 1, 621–625 (1980), 4-nitrodiphenylamine did not induce unscheduled DNA synthesis (UDS) or DNA replication in S-phase cells (no further details; Monsanto, 1986 d).

7.7 Carcinogenicity

No information available.

7.8 Reproductive toxicity

Pregnant Sprague-Dawley rats were treated with 0 (controls), 250, 1000 or 3000 mg 4-nitrodiphenylamine/kg body weight/day by gavage from days 6 to 15 of pregnancy. Each dose and control group consisted of 25 rats. Laparotomy was carried out on day 20 of pregnancy. No maternal- or embryotoxicity was observed at 250 mg/kg. In the

1000 and 3000 mg/kg dose groups, respectively, 1 (4%) and 8 (32%) of the dams died within the treatment period. Further indications of toxic effects on the dams included clinical signs of toxicity (no further details), a significant inhibition of body weight gain and a significant reduction in feed consumption from 1000 mg/kg body weight (level of significance not given). Statistically significant reductions in foetal weights were recorded in the 3000 mg/kg dose group, together with an increase in the number of offspring with 27 "presacral" vertebrae and slight, but not significant, reductions in uterus weights. A slight increase in the number of rudimentary ribs was observed after administration of 1000 and 3000 mg/kg. All other parameters, such as the numbers of corpora lutea, implantations/dam, resorptions, live foetuses per litter and pre- and post-implantation losses were within the range of the controls in all dose groups (Monsanto, 1991; Bannister et al., 1992).

7.9 Effects on the immune system

No information available.

7.10 Neurotoxicity

No information available.

7.11 Other effects

No information available.

8. Experience in humans

No cases of sensitization have been observed to date following contact with 4-nitrodiphenylamine (OECD SIDS, 1991).

9. Threshold limit values

No information available.

References

Bayer AG
DIN-Safety data sheet 4-Nitrodiphenylamin (1990)

Bayer AG
AIDA-Grunddatensatz Benzenamine, 4-nitro-N-phenyl (1991)

Bayer AG
Written communication to BG Chemie of 18.01.1993

Bannister, R.M., Brewster, D.W., Rodwell, D.E., Schroeder, R.E., Barnett, J.W., Jr.
Developmental toxicity studies in rats with 4-aminodiphenylamine (4-ADPA) and 4-nitrodiphenylamine (4-NDPA)
Toxicologist, 12 (1), 103 (1992)

CCR (Cytotest Cell Research GmbH & Co. KG)
Salmonella typhimurium reverse mutation assay with 4-nitrodiphenylamine, BALK No. 88/068
Unpublished report no. R4777 (1989)
On behalf of Bayer AG

Marhold, J.
Sbornik vysledku toxikologickeho vysetreni latek a pripravku, p. 134
Institut pro vychovu vedoucich pracovniku chemickeho prumyclu, Prague (1972)

Monsanto
SRI Project no. 8669 (undated)
Cited in: OECD SIDS (1991)

Monsanto
Product Assessment Report 4-NDPA (1979)
Cited in: OECD SIDS (1991)

Monsanto
Study ML-83–217 (1983a)
Cited in: OECD SIDS (1991)

Monsanto
Study BD-83–314 (1983b)
Cited in: OECD SIDS (1991)
See also: Bayer (1991)

Monsanto
Study BD-83–318 (1983c)
Cited in: OECD SIDS (1991)

Monsanto
Study PK-86–128 (1986a)
Cited in: OECD SIDS (1991)

Monsanto
Study PK-86–198 (1986b)
Cited in: OECD SIDS (1991)

Monsanto
Study PK-86–222 (1986c)
Cited in: OECD SIDS (1991)

Monsanto
Study SR-86–175 (1986 d)
Cited in: OECD SIDS (1991)

Monsanto
Rat teratology study (preliminary results) with 4-nitrodiphenylamine (4-NDPA)
Monsanto study no. SB-89–471 (1991)
NTIS/OTS 0530384

OECD SIDS Profile (Screening Information Data Set)
Benzenamine, 4-nitro-N-phenyl
OECD High Production Volume Chemicals Program (1991)

Randall, D.J., Bannister, R.M.
Acute toxicologic evaluation of 4-nitrodiphenylamine
J. Am. Coll. Toxicol. Part B,Acute Toxicity Data, 1, 66–67 (1990)

Sax, N.I., Lewis, R.J. (eds.)
Dangerous properties of industrial chemicals
7th ed., vol. III, p. 2516
Van Nostrand Reinhold, New York (1989)

Vogt, P.F., Gerulis, J.J.
Amines, aromatic
In: Ullmann's encyclopedia of industrial chemistry
5th ed., vol. A2, p. 37–55
VCH Verlagsgesellschaft, Weinheim (1985)

Weast, R.C. (ed.)
CRC Handbook of chemistry and physics
69th ed., p. C-424
CRC Press, Inc., Boca Raton, Florida (1988/89)

Ammonium persulfate

1. Summary and assessment

Ammonium persulfate is moderately toxic (LD_{50} rat oral 820 and 689 mg/kg body weight).

A no effect level of 1 mg/m^3 has been found in rats in a 7-day inhalation study with daily exposure for 23.5 h; body weights are reduced from 4 mg/m^3 and wet lung weight, lung protein and lung DNA content are significantly increased. Tracheal glycoprotein secretion (determined in tracheal explants) is also increased. The observed effects are interpreted as indications of inflammation and/or oedema formation.

In rabbits, ammonium persulfate in not irritating to the intact skin, but is mildly irritating to the eyes.

Ammonium persulfate is sensitizing to guinea pig skin.

No studies are available on the systemic effects of ammonium persulfate on repeated administration.

In view of the generally similar profile of effects of sodium and ammonium persulfate, the results of a 13-week feeding study with sodium persulfate in rats can presumably also be applied to ammonium persulfate, from which it can be inferred that administration of 300 mg/kg body weight/day will lead only to local damage to the mucous membrane in the gastro-intestinal tract; that systemic effects will not be observed; and that the administration of 30 mg/kg/day for 13 weeks or 100 mg/kg/day for 8 weeks with subsequent administration of 500 mg/kg/day for 5 weeks will be tolerated without adverse effects.

The available studies in rats and dogs given ammonium persulfate-treated flour or bread in the diet for 5 or 16 months, respectively, show no damage to or histopathological effects on the organs. No conclusions can be drawn on the subchronic or chronic toxicity of ammonium persulfate from these studies.

No indications of genotoxic activity of ammonium persulfate have been found in the Salmonella/microsome test or in a chromosome aberration test in vitro.

Epidermal hyperplasia developed in 2 animals in a 51-week skin-painting study in mice (20 females; ca. 1000 mg/kg body weight applied twice weekly), while no similar effects were observed in the acetone controls (15 mice). In a parallel promotion study (pretreatment with dimethyl benzanthracene), one epithelial carcinoma, one case of epidermal hyperplasia and one benign tumour were found. No hyperplastic changes were described in the control animals (15 mice, dimethyl benzanthracene plus acetone). Although the number of animals used was small and the study duration, at 51 weeks, was relatively short, the results of this study do not support a promotional or carcinogenic potential for ammonium persulfate, even at the high dose employed (ca. 1000 mg/kg body weight).

Compared with the histamine liberator 48/80, ammonium persulfate causes a lower liberation of histamine from rat and guinea pig skin sections and from rat mast cells (without destroying their membrane) in in vitro studies.

In humans, a large number of clinical observations have primarily described skin damage, but have also reported effects on the respiratory tract including asthmatic attacks and other intolerance reactions (unconsciousness).

An irritant effect has been held responsible for the reported skin effects seen following exposure to ammonium persulfate; however, clinical pictures of allergy are described considerably more often, occurring as both the immediate type with urticaria, pruritus, reddening and generalized oedema, and also as the delayed type with allergic dermatitis and chronic recurrent eczema.

The allergic reactions of the mucous membranes of the respiratory tract (rhinitis, asthma) should be regarded as being of the immediate type.

2. Name of substance

2.1	Usual name	Ammonium persulfate
2.2	IUPAC-name	Ammonium peroxydisulfate
2.3	CAS-No.	7727–54–0
2.4	EINECS-No.	231–786–5

3. Synonyms, common and trade names

Ammonium peroxysulfate
Ammoniumpersulfat
Diammonium peroxydisulfate
Diammonium peroxydisulphate
Diammonium persulfate
Peroxydisulfuric acid,
diammonium salt

4. Structural and molecular formulae

4.1 Structural formula

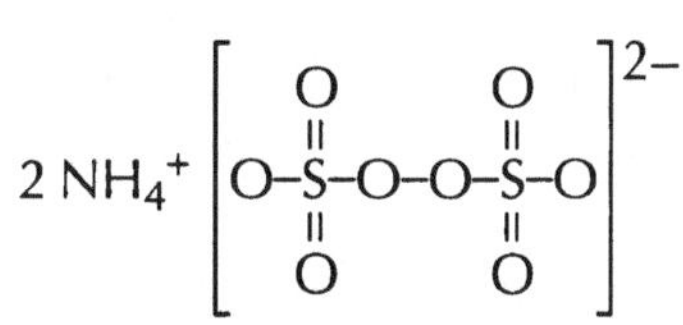

4.2 Molecular formula $N_2H_8O_8S_2$

5. Physical and chemical properties

5.1	Molecular mass, g/mol	228.2
5.2	Melting point, °C	120 (decomposition) (Neumüller, 1979)

5.3	Boiling point, °C	–
5.4	Vapour pressure, hPa	No information available
5.5	Density, g/cm^3	1.98 (Neumüller, 1979)
5.6	Solubility in water	560 g/l (at 20 °C) 740 g/l (at 50 °C) Decomposition begins at 60 °C (Degussa, 1988; Elektrochemische Werke München, undated)
5.7	Solubility in organic solvents	No information available
5.8	Solubility in fat	No information available
5.9	pH value	No information available
5.10	Conversion factor	1 ml/m^3 (ppm) ≙ 9.31 mg/m^3 1 mg/m^3 ≙ 0.11 ml/m^3 (ppm) (at 1013 hPa and 25 °C)

6. Uses

As a bleaching and oxidizing agent in cosmetics (e.g. hair lighteners), dyeing and fur finishing; as a reducer in photography; for pickling and dyeing surfaces and to etch printed circuits in the metal goods industry; as a polymerization catalyst in the plastics industry; as a disinfectant (Elektrochemische Werke München, undated; Neumüller, 1979).

7. Experimental results

7.1 Toxicokinetics and metabolism

No information available.

7.2 Acute and subacute toxicity

The oral LD_{50} value in male Wistar rats was 820 (530 to 1260) mg/kg body weight (14-day observation period; Smyth et al., 1969). A further LD_{50} value of 689 mg/kg has been reported in the rat (RTECS, 1993; ACGIH, 1986). Neither publication gave any information on signs of toxicity.

Groups of 6 male Sprague-Dawley rats inhaled ammonium persulfate aerosols at concentrations of approximately 1, 4, 9, 17 and 20 mg/m^3 for 7 days. Exposure was for 23.5 h daily and the average particle size was about 0.8 to 1.3 µm. For each exposure group there was a control group of the same size. In the ammonium persulfate-exposed rats, a dose-dependent loss of body weight (maximum 88%) occurred, which was significant at 4 mg/m^3 and above, as was the dose-dependent increase in wet lung weight (maximum 164%). No difference in the dry weight of the lungs was evident between groups. In contrast, total protein and total DNA content in the lungs were significantly increased at 4 mg/m^3 and above, without there being, however, an unequivocal dose response. Tracheal explants from all groups secreted more glycoprotein with the mucus, although only the value in the 4 mg/m^3 group was significantly increased. Overall, the results were interpreted as showing that the effects in the lungs were indicative of the test substance having inflammatory and/or oedema-causing potential, and the persulfate moiety was regarded as the actual pulmonary irritant. The no observed effect level was given as 1 mg/m^3 (Last et al., 1982).

7.3 Skin and mucous membrane effects

3 rabbits (Russian white) had 0.5 g of >99% pure ammonium persulfate moistened with 0.1 ml water applied to their clipped, intact or scarified skin under occlusive cover for 4 hours. Moderate to severe erythema, moderate oedema and the formation of scabs, and subsequently scars occurred after application to the scarified skin. In contrast, the intact skin showed only mild oedema, which cleared up within 24 h. The substance was thus not irritating to the intact skin (Asta, 1983 a).

Neat ammonium persulfate (0.1 g, >99% pure) caused severe diffuse reddening and swelling with hypersecretion when instilled into the conjunctival sac of the eye of 3 Russian white rabbits. The symptoms disappeared within 72 h, although clouding of the cornea was still present after 72 h. An irritation index of 10.5 was determined. Based on these results, the substance was evaluated as slightly irritating to the eye (Asta, 1983 b).

7.4 Sensitization

Ammonium persulfate (> 99% pure) was tested in 10 male and 10 female guinea pigs (Pirbright white, weight 310 to 468 g) in the optimization test following OECD guideline no. 406. 10 animals of each sex served as controls. On intradermal (0.1% solution in physiological saline) and dermal challenge (1% solution in demineralized water) 20/20 and 16/20 animals, respectively, showed positive reactions. The substance was thus sensitizing in guinea-pigs (Asta, 1985).

7.5 Subchronic and chronic toxicity

No subchronic studies on ammonium persulfate are available.

However, a 13-week study with *sodium persulfate* has been reported, the results of which can probably also generally be applied to ammonium persulfate. In this study, rats received sodium persulfate in the diet at doses of 30 or 300 mg/kg body weight/day for 13 weeks or 100 mg/kg body weight/day for 8 weeks followed by 500 mg/kg body weight/day for 5 weeks. No adverse effects on body weight, organ weight, haematological parameters, clinical chemistry values or urinary findings were detected in the treated rats. Histological evaluation revealed only epithelial damage to the gastro-intestinal tract at 300 mg/kg, possibly due to a local oxidative effect; other organs (no further details) were unaffected. No such epithelial damage was observed in the other treatment groups (Cascieri et al., 1981).

In connection with the study of ammonium persulfate as a flour additive, flour and bread were treated with 0.02 and 1% ammonium persulfate, respectively, and fed to rats for 5 months. The ammonium persulfate-treated flour and bread were added to the diet in "large amounts" (no exact details). No impairment of growth or of reproduction in the first or second generation (no precise details) were seen in the rats. Histopathological examination of "representative organs" of the offspring in both generations revealed no pathological effects (no further details; see also Sect. 7.8; Arnold and Goble, 1949).

In a further study, dogs (no details of breed or sex) were given a diet that contained flour treated with 0.15% (6 dogs) or 1% (4 dogs) ammonium persulfate for 16 months. 2 control animals were simultaneously given a diet containing a corresponding amount of untreated flour. "High levels" of the treated flour were added to the diet (no precise details). General condition, weight gain, phenolsulfophthalein clearance and blood picture were unaffected, as was the histopathological appearance of the kidneys, pancreas, gastro-intestinal tract, lymph nodes, thyroid gland, adrenal glands and CNS (Arnold, 1949; Arnold and Goble, 1950).

As reported in Sect. 7.7, the application of 0.2 ml of a solution containing 200 mg ammonium persulfate/ml acetone to the skin of 20 female Sencar mice twice weekly for 51 weeks (about 40 mg/mouse, ca. 1000 mg/kg at a weight of 40 g/mouse), had no effect on body weight gain or survival (no further details; Kurokawa et al., 1984).

7.6 Genotoxicity

7.6.1 In vitro

Ammonium persulfate (99.2% pure) dissolved in DMSO, was tested for mutagenic activity in the Salmonella/microsome test in the *Salmonella typhimurium* strains TA 92, TA 94, TA 98, TA 100, TA 1535 and TA 1537 with and without metabolic activation (S9-mix from the liver of Fischer rats pretreated with Kanechlor KC-400). Concentrations of up to 10 mg/plate were used (no further details). The substance was negative both with and without metabolic activation (Ishidate et al., 1984).

In a Salmonella/microsome test with pre-incubation (20 min at 37 °C), 98% pure ammonium persulfate dissolved in distilled water was negative in the *Salmonella typhimurium* strains TA 98, TA 100, TA 1535, TA 1537 and TA 1538 and in *Escherichia coli* WP2uvrA, with and without metabolic activation (S9-mix from the livers of KC-500 pretreated male Sprague-Dawley rats). Concentrations ranging from 1 to 5000 µg/plate were tested (Shimizu et al., 1985).

In a further Salmonella/microsome test with pre-incubation, ammonium persulfate dissolved in distilled water was tested for mutagenic activity at concentrations of between 0.1 and 10 mg/plate in the *Salmonella typhimurium* strains TA 97 and TA 102. The test substance was pre-incubated with S9-mix or phosphate buffer (pH 7.4) for 20 min. Ammonium persulfate was not mutagenic either with or without metabolic activation under these test conditions (Fujita and Sasaki, 1987; English abstract of a Japanese work).

Ammonium persulfate was cytotoxic but not mutagenic in the Salmonella/microsome test with pre-incubation (30 minutes in phosphate buffer pH 5.0) in the *Salmonella typhimurium* strain TA 97 without metabolic activation at 27 °C and 37 °C (no further details; Pagano et al., 1990).

99.2% pure ammonium persulfate dissolved in DMSO was tested in vitro in the chromosome aberration test in CHL cells (a cell line derived from Chinese hamster lung fibroblasts) at 3 different concentrations (maximum 250 µg/ml; no further details) without metabolic activation. The cells were fixed and stained after 24 and 48 h. After 48 h, the incidence of polyploid cells was 2%, while that of structural chromosome aberrations was 4% (no data on controls). Ammonium persulfate was not therefore mutagenic in this test system (Ishidate et al., 1984, 1988).

7.6.2 In vivo

No information available.

7.7 Carcinogenicity

20 female Sencar mice (6 weeks old at the beginning of the study) were treated with 0.2 ml of a solution containing 200 mg ammonium persulfate/ml acetone which was applied to their clipped skin twice weekly for 51 weeks (about 40 mg/mouse, ca. 1000 mg/kg at a weight of 40 g/mouse). After 51 weeks, 2 of the mice had epidermal hyperplasia (probably at the site of application), while the acetone controls

(15 mice) showed no effects of this type. In a second group of 20 female Sencar mice, dimethyl benzanthracene (DMBA; 20 nmol in 0.2 ml acetone) was applied to the skin one week before the beginning of treatment, and thereafter the mice were treated in the same way as the first group. During the 51-week treatment period, 1 mouse developed an epithelial cell carcinoma and a second developed epidermal hyperplasia; information on the third skin lesion which occurred was not provided; no tumours were found in the corresponding 15 control mice (DMBA plus acetone). Further information on local effects was not provided. Weight gain and survival relative to the controls were not affected in either study. Statistical evaluation of the skin effects in both studies revealed no significant differences compared with the controls (no further details; Kurokawa et al., 1984). Although the number of animals used was small and the study duration, at 51 weeks, was relatively short, the results of this study do not support a promotional or carcinogenic potential for ammonium persulfate, even at the high dose employed (ca. 1000 mg/kg body weight).

7.8 Reproductive toxicity

Rats were fed a diet containing "a high proportion" of flour or bread treated with 0.02 and 1% ammonium persulfate, respectively, for 6 months. Histopathological examination of "representative organs" of the offspring of the first and second generations revealed no effects (no precise details of study protocol or evaluation; Arnold and Goble, 1949).

7.9 Effects on the immune system

In order to study whether ammonium persulfate is able to release histamine, 500 µm-thick sections of abdominal skin from rhesus monkeys, guinea pigs (Dunkin-Hartley) and rats (GFY) were incubated with 1 to 1000 µg ammonium persulfate/ml in Tyrode solution for 15 to 30 min at 37 °C. None of these concentrations liberated histamine from the skin of the monkey or the guinea-pig. In the presence of 1000 µg/ml, 20 to 24% of the histamine contained in the preparation was liberated from the rat skin, an effect which was, however, characterized as a non-specific toxic reaction of the mast cells of the skin (no temperature dependence on cooling the preparation from 37 °C to 0 °C). The histamine liberator 48/80 (10 and 100 µg/ml) released histamine (up to 46%) from skin preparations from all three species (Mahzoon et al., 1977).

In a further study, 500 µm thick abdominal skin sections from the guinea pig were incubated in Tyrode solution with 200 to 16,000 µg/ml ammonium persulfate or with 200 µg/ml of compound 48/80 at 37 °C or 4 °C for 30 min. In this case, ammonium persulfate caused dose-dependent histamine liberation (up to 14.42% at 16,000 µg/ml, 37 °C), which was inhibited, but not prevented, at 4 °C (7.59% at 16,000 µg/ml). At 37 °C, 200 µg/ml of compound 48/80 led to the liberation of 10.3% of the available histamine, while at 4 °C, no more histamine was released (Parsons et al., 1979).

In a study of vascular permeability in guinea pig skin with the Evans-blue method (intradermal injection of ammonium persulfate at different concentrations in guinea-pigs that had been injected intravenously with Evans-blue 20 min previously), the effect of 16 mg ammonium persulfate/ml corresponded to that of 80 µg/ml of compound 48/80 with respect to the size and intensity of the area of skin stained blue (Parsons et al., 1979).

The same authors also investigated the effect of ammonium persulfate on isolated peritoneal mast cells from Wistar rats. Various concentrations of ammonium persulfate solution (100 µl, 0.33 to 2.7 µg/ml) were added to a mast cell suspension (10^5 cells/ml) which was then incubated for 40 min. After incubation, histamine was measured spectrofluorometrically in the supernatant. At 37 °C, ammonium persulfate caused dose-dependent histamine release of up to about 25% of the total histamine content. At 4 °C, histamine liberation was reduced, but not completely inhibited (ca. 7% of the total histamine content at 2.7 mg/ml). At the same concentration, ammonium persulfate did not cause haemolysis of sheep erythrocytes, from which it was concluded that the histamine release was not caused by destruction of the mast cell membrane, but by degranulation of the mast cells (Parsons et al., 1979).

7.10 Neurotoxicity

No information available.

7.11 Other effects

Myocytes from the hearts of male Sprague-Dawley rats (weight 250 to 300 g) were incubated with ammonium persulfate at concentrations of from 10 to 60 mM for up to 180 min at 22 °C. The cells were then suspended in 1 ml buffer containing 0.1 µCi (3.7 kBq) $^{45}CaCl_2$, and incubated for a further 5 to 20 min at 37 °C. Ammonium persulfate induced contraction of the myocytes, the number of contracted myocytes increasing with time. The concentration causing contraction of half of the cell population within 90 minutes was 55 mM. After incubation with 20 mM ammonium persulfate for 60 min, the ^{45}Ca content of the myocytes was significantly reduced relative to untreated cells (Kaminishi et al., 1989).

8. Experience in humans

Clinical effects on human skin seen after exposure to ammonium persulfate have been attributed sometimes to an irritant effect but in most cases to skin sensitization or allergic reactions.

The following observations indicative of an irritant effect have been described:

Reddening and slight crust formation on the skin of the head and forehead were observed in 2 people following the use of a hair-bleaching agent that contained ammonium persulfate. As no skin reactions were observed after testing with 2.5 and 5% ammonium persulfate (standard 24-/48- hour patch test), the reaction was attributed by the authors to an irritant effect rather than to sensitization (Fisher and Dooms-Goossens, 1976).

In workers in a factory manufacturing ammonium persulfate and potassium persulfate, skin rashes (itchy red papules and eczematous red patches on uncovered areas of skin) were observed in 20 to 70% of the newly employed workers. No patch tests were carried out in the affected subjects. However, the authors took the view that in most cases the skin lesions were a consequence of the irritant effect of the persulfates (White et al., 1982).

A concentration of 5% ammonium persulfate was described as primary irritant (Calnan and Shuster, 1963), while another report states that when 20 subjects with healthy skin were tested with ammonium persulfate at concentrations of from 5 to 10%, a questionably positive skin reaction occurred in only one person (Forck, 1968).

Sensitization and allergic effects:

2 workers in a plant manufacturing persulfates (sodium, potassium and ammonium persulfate) became ill for the first time about 2 weeks after taking up their posts, with dermatitis and then rhinitis, bronchitis and bronchial asthma. The causative role of the persulfates was confirmed by case history, skin tests, occupational exposure tests and by removal of the workers from their jobs. Delayed reactions were seen in both patients after skin tests with 1 and 5% solutions of sodium, potassium and ammonium persulfate. Studies at the workplace involving exposure for 8 hours led to a pathological increase in respiratory resistance, which disappeared within a few days of cessation of exposure. Following the dermal tests, histological examination of the skin revealed acanthosis, papillomatosis and parakeratosis of the epidermis with crusting and cell detritus, as well as persistent lymphocyte infiltration of the corium. Besides the allergic reactions, the authors also discussed chemical irritant and toxic effects (Baur et al., 1979).

Ammonium persulfate was tested in patients with contact dermatitis, eczema and generalized urticaria in standard 24- and 48-hour patch tests. Positive reactions were observed at concentrations of 2 to 5% (as aqueous solutions or in vaseline). The allergic reactions of the skin described in a large number of communications were seen as dermatitis, eczema, papules and generalized urticaria (Barsotti et al., 1951; Baur et al., 1979; Brubaker, 1972; Brun et al., 1966; Brun, 1975; Calnan and Shuster, 1963; Cronin, 1980; Fisher, 1982, 1985 a, b, 1989, 1993; Fisher and Dooms-Goossens, 1976; Gaultier et al., 1966; Guerra et al., 1992; van Joost et al., 1984; van Joost and Roesyanto, 1991; Kellett and Beck, 1985; Kleinhans and Ranneberg, 1989; von Krogh and Maibach, 1981; Marks and Cronin, 1977; Meindl and Meyer, 1969; Nava et al., 1983; Plunkett, 1976; Reiffers et al., 1974; Schulz, 1967; Widström, 1977; White et al., 1982; Young, 1974).

Most of these clinical observations have been made in the production of ammonium persulfate, in the hairdressing profession due to the use of ammonium persulfate-containing hair bleaches and in bakers who had contact with ammonium persulfate-containing flour (where it has been used as a bleaching agent; this flour additive has been banned in the Federal Republic of Germany since 1956; Forck, 1968). A particularly detailed description of ammonium persulfate allergy in bakers is given by Forck (1968) and in hairdressers by Kellett and Beck (1985).

In addition to urticaria and allergic dermatitis, bronchial asthma has also been seen, although relatively rarely, in the same subjects, particularly after contact with ammonium persulfate-containing bleaches in the hairdressing trade (Meindl and Meyer, 1969; Blandin, 1970; Baur et al., 1979; Pepys et al., 1976; Fisher and Dooms-Goossens, 1976; Gaultier et al., 1966; Blainey et al., 1986; Gamboa et al., 1989; Pankow et al., 1989; Subiza-Martin, 1951).

Ammonium persulfate occasionally provokes marked intolerance reactions immediately after exposure, with brief loss of consciousness and fatigue and the simultaneous occurrence of a generalized, itchy skin reaction; these effects were judged to be a histamine reaction (Calnan and Shuster, 1963; Fisher and Dooms-Goossens, 1976).

Whether in this case the antibodies fixed to the mast cells cause histamine liberation with the antigen by mast cell degranulation, or whether ammonium persulfate causes histamine release directly, as is seen in vitro in skin sections from guinea-pigs, is not yet clear.

9. Threshold limit values

The TLV value in the USA is 5 mg/m^3. Limit values of 2 and 1 mg/m^3 exist in Denmark and Great Britain, respectively (RTECS, 1993).

References

ACGIH (American Conference of Governmental Industrial Hygienists)
Documentation of threshold limit values and biological exposure indices
5th ed., p. 468 (1986)

Arnold, A.
Effect on dogs of flours treated with various improving agents
Cereal Chem., 26, 46–51 (1949)

Arnold, A., Goble, F.C.
Ammonium persulfate as a flour maturing agent
Fed. Proc., 8, 377 (1949)

Arnold, A., Goble, F.C.
Studies with dogs fed flour treated with ammonium persulfate
Cereal Chem., 27, 375–382 (1950)

Asta-Werke AG, Toxikologisches Institut Degussa-Asta
Bericht über die Prüfung der lokalen Reizwirkung von Ammoniumpersulfat (APS) nach einmaliger Applikation an der Haut des Kaninchens (Patch-Test)
Unpublished report no. Ind.-TOX.-205–82/83 (1983 a)
On behalf of Degussa AG

Asta-Werke AG, Toxikologisches Institut Degussa-Asta
Bericht über die Prüfung der Schleimhaut-reizenden Wirkung von Ammoniumpersulfat (APS) nach einmaliger Applikation am Auge des Kaninchens
Unpublished report no. Ind.-TOX.-206–82/83 (1983 b)
On behalf of Degussa AG

Asta-Werke AG, Toxikologie Degussa-Asta
Ammonium-Persulfat (APS) – Prüfung auf sensibilisierende Eigenschaften an der Haut des Meerschweinchens (Optimierungs-Test)
Unpublished report no. Ind.-TOX.-207–84/85 (1985)
On behalf of Degussa AG

Barsotti, M., Parmeggiani, L., Sassi, C.
Manifestazioni di asma bronchiale e di eczema negli operai addetti ai reparti di produzione dell'acqua ossigenata
Med. Lav., 42, 49–68 (1951)

Baur, X., Fruhmann, G., Liebe, von, V.
Persulfat-Asthma und Persulfat-Dermatitis bei zwei Industriearbeitern
Respiration, 38, 144–150 (1979)

Blainey, A.D., Ollier, S., Cundell, D., Smith, R.E., Davies, R.J.
Occupational asthma in a hairdressing salon
Thorax, 4, 42–50 (1986)

Blandin, G.
Désensibilisation chez des coiffeurs (laque et décolorant)
Rev. Fr. Allergol., 10, 327–331 (1970)

Brubaker, M.M.
Urticarial reaction to ammonium persulfate
Arch. Dermatol., 106, 413–414 (1972)

Brun, R., Jadassohn, W., Paillard, R.
Test épicutané avec réaction immédiate (persulfate d'ammonium)
Dermatologica, 133, 89–93 (1966)

Brun, R.
Epidemiology of contact dermatitis in Geneva (1000 cases)
Contact Dermatitis, 1, 214–217 (1975)

Calnan, C.D., Shuster, S.
Reactions to ammonium persulfate
Arch. Dermatol., 88, 812–815 (1963)

Cascieri, T., Fletcher, M.J., Weinberg, M.S.
A ninety day dietary feeding study of sodium persulfate in rats
Toxicologist, 1, 149, abstract 542 (1981)

Cronin, E.
Contact dermatitis, p. 126–185
Churchill Livingstone, Edinburgh, London and New York (1980)

Degussa AG
Grunddatensatz für Grobstoffe Ammoniumpersulfat (1988)

Elektrochemische Werke München AG
Data sheet, Ammoniumpersulfat EWM (undated)

Fisher, A.A.
Dermatitis of the hands from food additives
Cutis, 30, 304–319 (1982)

Fisher, A.A.
The persulfates: a triple threat
Cutis, 35, 520–526 (1985 a)

Fisher, A.A.
The persulfates: a triple threat. Part II: Occupational exposures
Cutis, 36, 25–27 (1985 b)

Fisher, A.A.
Dermatitis due to sulfites in home permanent preparations. Part II
Cutis, 44, 108–109 (1989)

Fisher, A.A.
Four flushers: topical agents producing facial flushing simulating the systemic variety
Cutis, 51, 225–227 (1993)

Fisher, A.A., Dooms-Goossens, A.
Persulfate hair bleach reactions. Cutaneous and respiratory manifestations
Arch. Dermatol., 112, 1407–1409 (1976)

Forck, G.
Vorkommen und Persistenz von Persulfatallergien
Berufsdermatosen, 16, 84–92 (1968)

Fujita, H., Sasaki, M.
Mutagenicity test of food additives with *Salmonella typhimurium* TA 97 and TA 102(II)
Kenkyu Nenpo – Tokyo – Toritsu Eisei Kenkyusho, 38, 423–430 (1987)

Gamboa, P.M., de la Cuesta, C.G., Garcia, B.E., Castillo, J.G., Oehling, A.
Late asthmatic reaction in a hairdresser, due to the inhalation of ammonium persulphate salts
Allergol. Immunopathol., 17, 109–111 (1989)

Gaultier, M., Gervais, P., Mellerio, F.
Deux causes d'asthme professionnel chez les coiffeurs: persulfate et soie
Arch. Mal. Prof., 27, 809–813 (1966)

Guerra, L., Tosti, A., Bardazzi, F., Pigatto, P., Lisi, P., Santucci, B., Valsecchi, R., Schena, D., Angelini, G., Sertoli, A., Ayala, F., Kokelj, F.
Contact dermatitis in hairdressers: the Italian experience
Contact Dermatitis, 26, 101–107 (1992)

Ishidate, M., jr., Sofuni, T., Yoshikawa, K., Hayashi, M., Nohmi, T., Sawada, M., Matsuoka, A.
Primary mutagenicity screening of food additives currently used in Japan
Food Chem. Toxicol., 22, 623–636 (1984)

Ishidate, M., jr., Harnois, M.C., Sofuni, T.
A comparative analysis of data on the clastogenicity of 951 chemical substances tested in mammalian cell cultures
Mutat. Res., 195, 151–213 (1988)

Joost, van, T., Stolz, E., Hoek, van der, J.C.S., Clermonts, E.
Shifts of delayed immune response to persulphates and other allergens
Contact Dermatitis, 11, 159–162 (1984)

Joost, van, T., Roesyanto, I.D.
Sensitization to persulphates in occupational and non-occupational hand dermatitis
Contact Dermatitis, 24, 376–378 (1991)

Kaminishi, K., Yanagishita, T., Kako, K.J.
Oxidant injury to isolated heart cells
Can. J. Cardiol., 5 (3), 168–174 (1989)

Kellett, J.K., Beck, M.H.
Ammonium persulphate sensitivity in hairdressers
Contact Dermatitis, 13, 26–28 (1985)

Kleinhans, D., Ranneberg, K.M.
Sofortreaktionen durch Ammoniumpersulfat in Blondierpulvern
Allergologie, 12 (8), 353–354 (1989)

Krogh, von, G., Maibach, H.I.
The contact urticaria syndrome – an updated review
J. Am. Acad. Dermatol., 5, 328–342 (1981)

Kurokawa, Y., Takamura, N., Matsushima, Y., Imazawa, T., Hayashi, Y.
Studies on the promoting and complete carcinogenic activities of some oxidizing chemicals in skin carcinogenesis
Cancer Lett., 24, 299–304 (1984)

Last, J.A., Dasgupta, P.K., DeCesare, K., Tarkington, B.K.
Inhalation toxicology of ammonium persulfate, an oxidant aerosol, in rats
Toxicol. Appl. Pharmacol., 63, 257–263 (1982)

Mahzoon, S., Yamamoto, S., Greaves, M.W.
Response of skin to ammonium persulphate
Acta Derm. Venereol. (Stockh.), 57, 125–126 (1977)

Marks, R., Cronin, E.
Hand eczema in hairdressers
Aust. J. Derm., 18, 123–126 (1977)

Meindl, K., Meyer, R.
Asthma und Urtikaria im Friseurberuf durch persulfat-haltige Blondiermittel
Zentralbl. Arbeitsmed., 3, 75–79 (1969)

Nava, C., Beretta, F., Elena, A., Ghizzi, A., Pattarin, R., Villa, L.
Le dermatiti allergiche da miglioranti e da altri additivi per farine
Med. Lav., 74, 376–379 (1983)

Neumüller, O.A. (ed.)
Römpps Chemie-Lexikon
8th ed., vol. 1, p. 191
Frankh'sche Verlagshandlung, Stuttgart (1979)

Pagano, D.A., Zeiger, E., Stark, A.A.
Autoxidation and mutagenicity of sodium bisulfite
Mutat. Res., 228, 89–96 (1990)

Pankow, W., Hein, H., Bittner, K., Wichert, von, P.
Persulfat-Asthma im Friseurberuf
Pneumologie, 43, 173–175 (1989)

Parsons, J.F., Goodwin, B.F.J., Safford, R.J.
Studies on the action of histamine release by persulphates
Food Cosmet. Toxicol., 17, 129–136 (1979)

Pepys, J., Hutchcroft, B.J., Breslin, A.B.X.
Asthma due to inhaled chemical agents - persulphate salts and henna in hairdressers
Clin. Allergy, 6, 399–404 (1976)

Plunkett, E.R.
Handbook of industrial toxicology
2nd ed., p. 26–27
Heyden, New York (1976)

Reiffers, J., Hunziker, N., Brun, R., Vidmar, B.
Sensibilisations cutanées allergiques peu communes
Dermatologica, 148, 285–291 (1974)

RTECS (Registry of Toxic Effects of Chemical Substances)
Peroxydisulfuric acid, diammonium salt, SEO350000
produced by NIOSH (National Institute for Occupational Safety and Health) (1993)

Schulz, K.H.
Berufsdermatosen - ausgewählte Kapitel
Z. Haut Geschlechtskr., 42, 499–509 (1967)

Shimizu, H., Suzuki, Y., Takemura, N., Goto, S., Matsushita, H.
The results of microbial mutation test for forty-three industrial chemicals
Jpn. J. Ind. Health, 27, 400–421 (1985)

Smyth, H.F., jr., Carpenter, C.P., Weil, C.S., Pozzani, U.C., Striegel, J.A., Nycum, J.S.

Range-finding toxicity data: list VII
Am. Ind. Hyg. Assoc. J., 30, 470–476 (1969)

Subiza-Martin, E.
Eczema y asma producidos por la asociacion de harina y persulfato amonico
Alergologia, 5, 343–348 (1951)

White, I.R., Catchpole, H.E., Rycroft, R.J.G.
Rashes amongst persulphate workers
Contact Dermatitis, 8, 168–172 (1982)

Widström, L.
Allergic reactions to ammonium persulphate in hair bleach
Contact Dermatitis, 3, 343 (1977)

Young, E.
Allergic reactions among bakers
Dermatologica (Basel), 148, 39–42 (1974)

Compound chemical index vol. 1-10

MIX
Papier aus verantwortungsvollen Quellen
Paper from responsible sources
FSC® C105338

FSC
www.fsc.org

If you have any concerns about our products,
you can contact us on
ProductSafety@springernature.com

In case Publisher is established outside the EU,
the EU authorized representative is:
Springer Nature Customer Service Center GmbH
Europaplatz 3, 69115 Heidelberg, Germany

Printed by Libri Plureos GmbH
in Hamburg, Germany